Heaven's Touch

✦

James B. Kaler

Heaven's Touch

✦

From Killer Stars
to the Seeds of Life,
How We Are Connected
to the Universe

PRINCETON UNIVERSITY PRESS   PRINCETON AND OXFORD

Published by Princeton University Press, 41 William Street, Princeton, New Jersey 08540
In the United Kingdom: Princeton University Press, 6 Oxford Street, Woodstock,
Oxfordshire OX20 1TW

ISBN: 978-0-691-12946-4

Library of Congress Control Number: 2009924763

British Library Cataloging-in-Publication Data is available

This book has been composed in Sabon with ITC Benguiat Book display

Printed on acid-free paper ∞

press.princeton.edu

Printed in the United States of America

1 3 5 7 9 10 8 6 4 2

To Maxine and all my family, with love

.✦. **Contents**

⋆ **Preface and Acknowledgments**

Most books are about "something," a specific topic. In my own context of astronomy, about "stars" or "galaxies" or "cosmology." This one seems to be about "everything," everything I know that would be relevant to how we are directly connected to the heavens, literally to the entire Universe. Hardly alone and isolated, we—the Earth, humanity—are under the constant influence of the cosmos, from tides to solar magnetism to asteroid collisions to "killer stars." While some of these influences are the stuff of cosmic disasters (and bad movies), most are in fact not just benign, but necessary for our creation and continued existence. Here we look at both aspects, which are inexorably linked together, wherein we take a tour from Earth and Moon, through the Solar System, and out to the far reaches of space, all of which touch us in a remarkable variety of ways.

No book is created in a vacuum. I'd particularly like to thank my editor, Ingrid Gnerlich, for her belief in the project and her continued advice and help, and my chief science advisor, Brian Fields, for leading me though some of the intricacies of cosmic rays, supernovae, and cosmology. Special thanks also go to Ron Webbink for his instruction, to the three anonymous reviewers who made numerous critical corrections, and to all the people and institutions who graciously helped supply the book's images.

Those institutions identified in the illustration captions by abbreviation are AURA (Association of Universities for Research in Astronomy), BATSE (Burst and Transit Source Experiment), CGRO (Compton Gamma Ray Observatory), CXC (Chandra X-Ray Observatory Center), HST (Hubble Space Telescope), ISAS (Japanese Institute for Space and Aeronautical Science), ESA (European Space Agency), JHU/APL (Johns Hopkins University Applied Physics

Lab), JPL (Jet Propulsion Lab), NASA (National Aeronautics and Space Administration), NOAO (National Optical Astronomy Observatory), NSF (National Science Foundation), SOHO (Solar and Heliospheric Observatory), STScI (Space Telescope Science Institute), the WIYN (Wisconsin, Indiana, Yale, and NOAO) Observatory, and WMAP (Wilkinson Microwave Anisotropy Probe).

My gratitude as well goes to production editor Brigitte Pelner, copy editor Karen Verde, designer Leslie Flis, and illustration specialists Dimitri Karetnikov and Erin Suydam for their fine work.

At the top of the list, however, are the generations of scientists and writers who mastered the art of astronomy and who discovered what you will find within. Thanks finally, as always, to Maxine Kaler for her continued support.

Please enjoy the tour. And watch out for rising tides and falling asteroids.

Jim Kaler, Urbana Illinois

Heaven's Touch

✦

✦. Chapter 1

Reaching Out

arth. A beautiful word, Earth. It summons visions of mountains, oceans, blue skies and clouds, wind and rain, prairies and plowed fields, life. Our small planet is utterly central to us. Whatever happens almost anywhere on it affects our lives. The only body of comparable importance is the Sun, which provides heat, light, nearly all of our energy. Without either Sun or Earth, we obviously could never have come to be. We are the sons and daughters of both.

The nighttime sky reveals a different face. The blue cover is replaced by a black one alight with stars, shifting planets, the Moon, and occasional comets and meteors. Though foundations of art, music, philosophy, the residents of night seem otherwise to have little effect. Looking outward from our home, we feel suitably isolated. We might spend a lifetime watching the heavens, yet nothing much, if anything at all, seems to happen there that has anything to do with our lives.

But that is only because we do not look closely enough, or live long enough, to witness the whole story. While some influences have long been known (lunar tides, for example), only over the past century or so have we learned that we are profoundly affected not just by the Sun and Moon, but by practically everything that happens "out there." We are not just the children of Earth and Sun, but of the starry Universe.

Among the most amazing discoveries of modern astronomy is that even our day-to-day affairs are in fact subject to the vagaries of distant planets and stars. No astrology here: just good science that has uncovered the true planetary and stellar influences, which are far grander than any "magical" ones could ever be. Over the nine chapters that follow, they will be revealed as we reach out from our home ever deeper into the heart of outer space to experience not isolation, but a grand synergy in which everything influences everything else. But first we need a summary of what is actually out there, so as to provide a context for what is to come.

Stars

Stars surround us, eight to ten thousand of them visible to the eye alone. A handful brilliantly punctuate the darkness, while the fainter ones overwhelm our vision with their sheer numbers. There is no record of discovery. Stars have been seen, admired, loved, feared, studied, romanticized, since humans first looked upward. Long shrouded in mystery, they were believed by our ancestors to have been placed by the gods into patterns—constellations—to tell stories, to instruct, to commemorate, to note the passage of time. To the north roam the two celestial Bears, Ursa Major and Minor; to the south stalks Orion the Hunter, accompanied by his two canine companions, Canis Major and Minor. Farther south sails the ship of the Argonauts, while girdling the sky in a great circle are the twelve ancient animalistic figures of the Zodiac, which cradle the Sun (Libra once being the Scorpion's claws).

All cultures recognized such stellar patterns, "ours" coming down to us from Babylonia and before. Ultimately, forty-eight of them were passed to us through the hands of the ancient Greeks and Arabs. Thousands of years after their invention, we still celebrate them from our own backyards, allowing our forebears to reach out to touch us across the ages. These ancient constellations are supplemented by new ones culled from the interests of our own more modern times (a furnace, sextant, microscope), giving us eighty-eight of all sizes and shapes displayed across the northern and southern celestial hemispheres, to which are added dozens of informal configurations.

Telescopes reveal an uncountable population of stars: millions, billions. In reality they are other suns of all colors, kinds, forms, sizes, and ages. Self-luminous, they send us signals of light from

Figure 1.1. The bright half of the Milky Way, the combined light of the stars in the disk of our Galaxy, spills across the sky from north at left to south at right. The complex dark bands are made of thick clouds of interstellar dust that are the birthplaces of stars. The Southern Cross lies at the upper right-hand corner. The left-hand star of the pair just down and to the left of the Cross is Alpha Centauri, the closest star to Earth, four light-years away. Antares in Scorpius is the bright star just above center. Courtesy of Serge Brunier.

distances measured in impossibly long units. The Sun, 150 million kilometers (nearly 100 million miles) away, provides a fundamental measure. One hundred times the diameter of Earth, our star holds more than 300,000 times Earth's mass. Running on nuclear power (the conversion of hydrogen into helium), one second's worth of its radiation could provide all the energy used on Earth for the next million years. Other stars range in size from that of a small city to the orbits of the giant planets; in mass from a few percent that of the Sun to over 100 times solar; in age from newly born to nearly the age of the Universe itself. Once created, they live mostly quiet lives until their internal fuel runs out, at which point they enter desperate straits, first swelling to gigantic proportions before dying as tiny, worn-out cinders. Along the way they

produce some of the most beautiful sights of nature. A tiny few
even explode, whence in their nuclear fury they manufacture most
of the chemical elements of which we are made.

Planets

Against the distant stellar background lie the planets of our Solar
System. All but the farthest of them appear to outshine the stars,
which are hundreds of thousands, millions, of times farther away.
As the planets orbit the Sun, each on its own path, each taking its
own time, their movements against the Zodiac long ago captured
the imagination, so much so that in ancient times they were related
to gods. There flies fleet Mercury the Messenger. Now you see him in
evening twilight, but look quickly before he flits into morning. Visit
next with peaceful Venus, the classic brilliant evening or morning
star, Aphrodite, goddess of love and beauty, whose glow can be seen
in full daylight and can cast eerie shadows at night. Jumping over
Earth, find reddish Mars, the leering god of war, half again farther
from the Sun than we. Then walk with stately Jupiter, king Zeus
himself. Five times Earth's distance from the Sun, he spends a year
visiting each of the dozen zodiacal figures, every 20 years passing
his doubly distant, slower, fainter, defeated father, Saturn. Perhaps
from their zodiacal homes the gods can tell us the meanings of our
lives, can reveal our fates, the fortune-telling art of astrology not
splitting off from astronomy until we began to see the planets for
what they actually are: other "earths" cast into a variety of forms.
To these, add the two discovered in modern times. Another near-
doubling of distance gets us to Uranus, named after the embodi-
ment of the heavens themselves. Half again farther out gets us to
Neptune, named for the god of the Sea. Discovered only in 1846, it
takes a century and a half to make a full loop of the Sun.

Nearby, orbiting our Earth and brighter still, is our small sister
planet, the Moon, which constantly changes shape as she makes
her monthly rounds. Not only can we see the Moon resolved as a
disk with the naked eye, it's so close that we can even see dark features on its surface, of which the old-timers made fanciful figures,

but which we now know are ancient lava flows closely related to a battered, beaten, punctured, cratered surface. In between is the leftover debris of the formation of our planetary system, made of asteroids that flock between Mars and Jupiter and of the icy comets that thrive mostly beyond Neptune and that are related to the "last planet," tiny Pluto. While rare among the small planets of the inner Solar System, dozens of moons flock around the giant planets from Jupiter on out.

While the interactions between the stars, Sun, Moon, and planets with our Earth are deep and complex, our first appreciation of them all is still through the light they send us. Stars and the Sun make their own radiant energy, while the planets and the Moon shine by reflection—as does the Earth. However it is produced, our first thought is always to look to the

Light.

Light: our human window to our surroundings and to the sky. Nothing is so fast! Indeed, nothing *can* be so fast. Turn on an imaginary flashlight and three-billionths of a second later the radiant beam is a meter away; in a second and a half it has passed the Moon (384,000 kilometers, 239,000 miles), in eight minutes the Sun (150 or 93 million). In three hours it begins to exit the planetary system, zipping by Neptune at 300,000 kilometers (186,000 miles) per second, then an hour later past Pluto's path. Such a beam would next go on a long lonely ride for at least *four years* before it would encounter another star, one then said to be four "light-years" away. And it would fly for thousands of years before it passed the last star visible without a telescope.

Light: our human window to the past. Reverse the flow and let natural starlight fall toward you. Because of the time needed for light to travel, we see the most distant stars as they were thousands of years ago, Pluto as it was four hours ago. Sunlight is eight minutes old. Even your friends appear as they were a few billionths of a second ago. The "present" is as we see it with our own eyes. Everyone thus has a different view of reality.

Light: strange stuff. As a flow of alternating electric and magnetic fields (the two forever linked together, each causing the other), no one has ever actually "seen" light itself; we instead sense the effect it has on our eyes. A major means of moving energy from one place in the Universe to another, light behaves in part like a continuous wave, much like the wave patterns that glide across the sea. The shorter the distance between wave crests (the "wavelength"), the more waves hit you per second, and the more energy a wave train can carry. But light also acts like a pack of speeding bullets, as particles, "photons" that hurdle along from source of the light to the eye. Light's strangeness is that it behaves as waves and particles *at the same time,* a concept that renders our best intuitive imaginations powerless.

Another strangeness (to us, not to Nature) is that photons have no mass, no "weight"; they are the only particles known that do not. In his relativity theory, Einstein showed us that mass (M) and energy (E) are related through perhaps the most famous equation ever written, $E = Mc^2$, where c is the velocity of light. While the speed of light alone is huge, squaring it (multiplying it by itself) makes the number vastly larger, such that a tiny amount of mass can be turned into a startling amount of energy—which is the key to nuclear and stellar power.

Energy, which comes in many forms, can in its crudest sense be thought of as the ability of a body to accelerate or to give heat to another. With more mass and higher velocity, a speeding car obviously has more energy than a running human. Einstein also then revealed that as the velocity (hence energy) of a particle increases ("relative" to us), so does its perceived mass. At light-speed a particle's mass would become infinite, which is impossible. However, since photons have no mass to begin with, only they are allowed to run at the limit. Anything with mass is confined to less than light-speed.

For all its richness, our personal window on the Universe is terribly small within a stunning range of wavelengths, within the "electromagnetic spectrum." With our eyes, we see those waves that fall between 0.00004 and 0.00008 of a centimeter (where, not so oddly, the Sun and stars generally emit the most of their energy). We sense the different visible wavelengths as different colors. At the long end, we see red, at the short end violet, with orange, yellow, green, blue, and their hundreds of overlapping shades in between.

Outside of this visual band, our eyes cannot register wave-photons, no matter how powerful or how many there may be. Longer than the visual wavelength limit—up to about a millimeter—lies the "infrared." Longer waves, into kilometer-wavelengths toward an unknown end, we loosely call "radio." Conceptually, however, all are the same. All are still "light" that carries energy, all running (in the vacuum) at the "speed of light." Long waves mean low energy, such that (unless at high intensity or at specific wavelengths such as those found bouncing within a microwave oven) they pose little danger to us or other living things.

Shorter than the visual limit, more violet than violet, is the *ultraviolet*. If less than a percent or so of the wavelength of visual light, the waves are called X-rays. Another factor of 100 smaller, we enter the domain of "gamma rays." Short waves carry higher energies, resulting in increasingly higher danger. Ultraviolet light from the Sun will burn your skin, while X-rays (unless used properly for their medical benefit) are downright dangerous; gamma rays can be lethal. Fortunately, the latter two and most of the ultraviolet realm are blocked by the Earth's protecting atmosphere.

Together, stars, the dusty gases in the space between the stars, and related cosmic objects, emit across this entire wavelength array. Among the great triumphs of twentieth-century astronomy was the opening of the electromagnetic spectrum to our view via a startling array of new technologies. The expansion of our vision began in the 1930s with radio astronomy, and ended with gamma rays and X-rays observed from satellites orbiting in the depths of space above the surrounding atmosphere.

The visual spectrum from violet to red is but one octave on an imaginary electromagnetic piano with a keyboard hundreds of kilometers long. All of it is available for inspection, allowing us to explore the depths of the system of stars in which we live, of our Galaxy.

Figure 1.2. The Hubble Space Telescope orbits against the background of Earth and its clouds. Launched in 1990, the HST epitomizes the dozens of space observatories that can "see" with exquisite detail across the electromagnetic spectrum. NASA.

The Milky Way

Imagine a scene not from Earth's surface, where our planet creates a horizon that cuts out half the sky, but from deep space outside the Solar System, where you can get a full view of the celestial sphere that seems to surround us, and thus of all the stars that make themselves known to the unaided eye. No longer twinkling from the disturbing effects of Earth's atmosphere, they are point-like jewels set on a black cosmic cloth. Splitting the heavens in two is a luminous, highly irregular band of light—the Milky Way— made of the combined light of billions of stars that are individu- ally invisible. Toward the zodiacal constellation of Sagittarius (the Archer), the Milky Way is bright and broad, while in the opposite direction, toward Taurus (the Bull) and Auriga (the Charioteer), it appears faint and indistinct.

The Milky Way, the subject of countless tales and poems, is the visual manifestation of our Galaxy, our own real home, a collection of over 200 billion stars of which our Sun is just one. Most of the Galaxy's stars are arranged in a thin disk that is filled in the middle with an even thinner layer of free gas and dust. The system is so big that it would take a light ray more than 100,000 years to make the journey from one side to the other. There really is no definable visual edge: our stellar system just gradually fades away. We, set within the disk, see it around our heads as the storied band of light, while the layer of dusty gas clouds (from which new stars continuously condense) divides the band in two, like the filling of a cake. Though located well within the disk, we are far from the center, hence there is a dramatic variation in brightness as we gaze around the Milky circle. Surrounding the whole affair is a vast and encompassing, though sparsely populated, spherical halo.

Within the disk some stars gang together into clusters that tell of common birth. Among the most beloved of celestial sights are the Pleiades—the Seven Sisters—and their mythological half-sisters, the Hyades, both in residence in Taurus. Under the curve of the Larger Bear's Big Dipper flows another, Coma Berenices (Berenices's Hair), while off in Cancer's direction the Beehive buzzes with its delightful swarm. The telescope shows such clusters to contain hundreds, even thousands, of members, while in the Galactic halo are a handful of gigantic clusters that can contain millions of stars, some of their fuzzy glows also visible to the naked eye. Even if—perhaps like the Sun—stars have drifted away from their birthmates, they are still companionable, a goodly fraction of them double, triple, quadruple, even more, in a hierarchy of gravity-bound neighbors.

To keep the Galaxy from collapsing on itself through the combined gravity of its stars, the stars must be circulating—orbiting—causing our Galaxy to rotate, though not as a solid body. Instead, the inner portion rotates in much less time than the outer regions. Our own Sun takes 250 million years to make a full circuit. All the stars around us have their own unique paths that depend on

Figure 1.3. No, not the Dipper, but the Pleiades (Seven Sisters) star cluster that
graces the constellation Taurus. While only six stars are visible to the ordinary
eye, the cluster—430 light-years away—contains hundreds of fainter ones.
Bound together by gravity, the Pleiades is but one of thousands of clusters,
some of which contain millions of stars. The cluster's hottest and brightest stars
illuminate a thin dusty gas cloud through which the cluster is now passing.
Courtesy of Mark Killion.

the gravities they feel, such that all seem to drift past us or we past
them. Over the next thousands of millennia, the constellations will
slowly change their shapes until they become unrecognizable to
our current eyes, as the Sun and its Solar System travel on a slow
but steady journey to the Galaxy's other side. By then, none of
the stars we now see will be visible, but will instead be replaced
by others that we in our present era cannot see at all. Gravitational disturbances and rotation also help spread out the stars into
a huge pinwheel of graceful "spiral arms," rendering ours a classic
"spiral galaxy."

And if this system, our Galaxy, seems almost overwhelming,
there is more: we are not alone.

Figure 1.4. Want to see what our Galaxy might look like? The Andromeda galaxy is (like our own) a flat disk over 100,000 light-years across that lies some 2.5 million light-years away. New hot blue stars crowd outer spiral arms, while ancient reddish stars flock toward the center. Below center and to the upper right are a pair of small companion elliptical galaxies. NOAO/AURA/NSF/WIYN.

Galaxies

Off in vaster distances lie other galaxies. Four are visible to the naked eye. From the Earth's southern hemisphere you can easily spot a pair of small ones some 180,000 light-years away, the two irregularly shaped Magellanic Clouds, which orbit our larger Galaxy as satellites. In northern autumn evenings look to the constellation Andromeda to find a small cottony patch nicely visible without the telescope. The most distant thing the unaided human eye can see, the Andromeda Galaxy lies two and a half million light-years off. Another spiral galaxy even grander than our own, it may contain twice as many stars. Nearby, in the neighboring constellation of Triangulum (the eponymous Triangle), and at a similar distance, lies a much smaller spiral available only to younger eyes viewing from extremely dark sites.

Huge as these distances seem, they are just a small step into the cosmos. Ours, these four galaxies, plus another few dozen small scrappy ones are tied together through their own mutual gravities into a poorish cluster called simply "The Local Group." Farther out, 150 million light-years away, is a much grander cluster. Nearly filling the large zodiacal constellation Virgo, the Virgo Cluster contains thousands of galaxies, the biggest an ellipsoidal, nonspiral structure more than ten times as massive as our own Milky Way system.

The farther away we look, into distances measured in billions of light-years, the more we see—individual galaxies plus hundreds, thousands of their clusters, each containing thousands of individual members. The number of galaxies now seems almost infinite, one piling on top of the other, more galaxies than there are stars within the Milky Way. If we could add up all we could potentially see in our best telescopes, the number would approach a trillion galaxies, each one different, the larger ones containing billions, hundreds of billions, of stars, the most distant seen as it appeared billions of years in the past.

All—from our very selves, to our planets, to the nightly stars, to the Galaxy, to other galaxies—are locked together as part of the greatest of all structures, one that includes everything.

The Universe

In a kind of overwhelming simplicity, the Universe is "all there is." We have no idea how "big" it is, if that concept has any meaning at all. It might be infinite. However large, it has one great and singular characteristic: expansion. For nearly a century, we've known that the farther a galaxy is from us, the redder its light. The degree of this "redshift" is interpreted in terms of speed of recession: double the distance, double the velocity, and so on. It would seem that we are at some kind of center, with all the other galaxies madly fleeing from us. That view is an illusion. Since the velocities are in direct proportion to distance, every galaxy is moving away from every other galaxy. No matter in which one you might live, you would see the same thing.

At the same time, we cannot ignore the powerful effects of gravity. The expansion involves only large scales. The Earth is not expanding, and neither is the Galaxy; the mutual gravity of its components overwhelms any expansion effect. If galaxies are near each other, like ours and Andromeda, their mutual gravity is still strong enough to overcome the expansion, and keeps them together. Likewise, gravity maintains the integrity of even large clusters of galaxies. It is more accurate to say that isolated galaxies and clusters of galaxies are separating at velocities proportional to the distances between them.

The Universe then looks as if it has undergone some kind of gigantic explosion, with galaxies and their clusters hurling themselves through space. Another illusion. Einstein revealed an extraordinary depth to Nature, in which time and space are combined into a single four-dimensional entity called "spacetime." The apparent recession of galaxies is the result not of their movement through space, but of the expansion of spacetime itself. Galaxies, caught in its web, are merely going along for the ride.

The redshift is commonly misinterpreted as a "Doppler shift," which causes the light waves of a receding body to seem longer and thus redder, those of an approaching body to appear shorter and bluer. We hear the same thing in audio: an approaching car has a higher-pitched sound than a receding vehicle. Doppler shifts in the spectra of stars and other objects are critical to our understanding of them and of the Galaxy. What is really happening is that the photons from distant galaxies are stretched along with spacetime's expansion and therefore naturally lengthen, or "redden," during their long journeys to our eyes. The expansion rate gives the age since the beginning of the still-mysterious expansion. We—in the broadest sense—were born 13.7 billion years ago at a moment when all matter and energy were crushed into a "hot, dense state" whose actual origin is completely unknown. Our Sun and its planets, born from condensing dusty gases of interstellar space, came along 9 billion years later.

The evidence that the creation event, the "Big Bang," actually took place is deep and rich. Among the best is that the "fireball" from the Big Bang produced not only the mass of the Universe but—from the intense energy of its superheated state—a flood of short-wave, high-energy gamma rays. Just like the light from distant galaxies, over the past billions of years, they were stretched, becoming longer, that is, "redder." Another way of looking at it is that the heat of the original early Universe cooled such that it now contains only low-energy radio waves. We live in a constant bath of them called the Cosmic Microwave Background, whose radio photons surround us as if radiated from a body at a mere three Celsius degrees above the lowest possible temperature (absolute zero, –273 C), which is the value predicted by Big Bang theory. Ripples in this background represent the beginnings of assemblies of galaxies and also give us the same age as found from the galaxies' velocities. And nearly 14 billion years later, here we are, here are the stars, here are the planets, here are the heavens themselves.

Darkness

Yet for all our seeming knowledge, as did the stars of old, the Universe and its treasures remain shrouded in mystery. We have done a sort of inventory of the stores of energy and mass that it contains. The key to understanding lies in the Universe's "shape." Spacetime is rather like a four-dimensional sheet of rubber (however unimaginable that might be) that might be bent, distorted, even folded back upon itself in a variety of ways that depend critically upon the average amount of matter contained within a specified huge volume. Shape is crucial to know, as the calculated age of the Universe depends on it. The practitioners of the art of "cosmology," the study of the cosmos at large, can calculate the average,

Figure 1.5. The Hubble "Ultra-Deep Field" reveals thousands of galaxies within a pinhead of the sky a mere twentieth of a degree across, a tenth the angular size of the full Moon. A few big bright ones are relatively nearby, while the faintest stretch out to billions of light-years away. The farther they are, the faster they move away from us, the first clue that the Universe was created in a "Big Bang." Though their number seems overwhelming, galaxies such as these represent less than 1 percent of the mass-energy of the Universe. NASA, ESA, R. Windhorst (Arizona State University), H. Yan (Spitzer Science Center, Caltech).

smeared out, density of mass (including the mass equivalent of energy) that is required to roll out spacetime with no bends whatever, to make it, in a weird multidimensional way, "flat" (such that Euclid's famed plane geometry actually works).

Add up the stars and the stuff in the spaces between them, and it totals to just under a percent of that needed. More mass is indicated by Big Bang theory, which predicts how primitive hydrogen atoms combined to create a variety of light chemical elements. Much, if not most, of this matter seems to be in the form of hot gas that lies between the galaxies of large clusters, and even between clusters, where it is observed with X-ray telescopes. Some is matter left over from galaxy formation, while much was also ejected from the cluster's members by stellar explosions. Or so it's thought. Summarizing the various predictions and observations gets us close to 5 percent of the flatness requirement.

Then descends the shroud. Our Galaxy, its stars revolving around the center under the influence of their combined gravity, is spinning too fast for what we see. Galaxies in clusters orbit around the cluster's centers under the influence of *their* mutual gravities, but again, they move faster than expected. There must be something out there with enough of a gravitational hold to do the job, to speed things up, but it is completely unseen. *Dark matter*. It surrounds galaxies, pervades their clusters. We have no idea what constitutes it. Rather, there are *many* ideas, but none that can be proven. Add it all up based on the amount of mass needed to yield dark matter's gravity, and lo, one finds another 20 percent, getting us up (once the numbers are rounded off) to a quarter of that required to unfold the Universe, but still not enough.

To resolve the issue (and to add to the mystery at the same time), look deeper at the expansion. Modern telescopes, imagers, and above all, knowledge, have allowed cosmologists to measure the expansion rate to distances of billions of light-years, allowing us to look far back into time. We might expect a slowdown of the expansion as the combined gravity of everything in the Universe acts to hold it back. Velocity and redshift are indeed seen *not* to be quite in direct proportion to each other. Instead of showing a slowdown, the observed variance reveals the opposite, a surprising speedup! The expansion is getting ever faster. Any acceleration requires energy, which from its mass equivalent yields the missing 75 percent. Within the uncertainties of the data, the Universe is flat! But we have even less of an idea where this energy comes from than we do about the nature of dark matter. So to dark matter, add *dark energy*. Either that or something is terribly amiss with our concept of gravity.

We seem almost to be back to primitive times when we first looked up to the stars to wonder what they are. We are still wondering just as hard, perhaps not about stars as such, but about the stuff that surrounds them, which surely has a role in making them, and that then plays a role in making us. So we keep wondering and exploring. And as we reach out in our attempts to understand them all, from planets to galaxies, they in turn reach back to us to aid in our quest.

Heaven's Touch

Return to the outpost in deep space from which we observed the Milky Way. The scenery is quiet, serene, strikingly beautiful. Nothing seems to happen. Floating in the void we feel isolated, just as we would back on Earth where our own planet dominates us and our thoughts. What do these stars have do with us? We cannot reach them. Their great distances make it likely that no spaceship ever will. And there is no evidence that anyone out there—if there is anyone—has ever, or will ever, come here. All the connections we have with the distant cosmos seem to be from beautiful and benign starlight, which allows us to admire and study the Universe, but never to reach out and touch it.

Similar concepts of alienation extend to near space, to our Solar System. The Sun warms us with its light and heat, but otherwise we pay it little heed. One hundred million miles away—a hundred times the solar diameter, ten thousand times Earth's diameter—it seems otherwise not to affect or bother us. The Moon and planets, going through their once-mystifying movements (which we now fully understand from gravitational and orbital theory), are pleasing and fun to watch, but again they appear to have no direct bearing upon our own small world. Though we have gone to them, imaged them up close, have even landed upon them out of curiosity and a yearning for exploration, we could as easily have left them alone, as they do us.

Or do they? The story of discovery over the past century or more has starkly revealed that such isolationist views are the reverse of reality. Our senses are limited, and our life on Earth short compared with the flow of celestial time, both of which veil many of the remarkably varied ways in which the heavens directly interact with us. Indeed, life itself would be impossible without all these direct interactions.

Begin by standing at the beach to watch the water go up and down in synchrony with the position and phase of the Moon, all powered by gravity from a satellite a quarter million miles away, the cycle modified by the far more distant Sun. Not just for the watching, tides are a part of the fabric of life, perhaps even in part responsible for its creation. They—the tides—are on Earth, but not of Earth, as their production lies in the heavens, which reaches out to touch our world.

Through the tides, the Sun barely reveals its power. Northerners will tell you more, about the shining lights in the night sky, the northern lights that hang, that flow, that shoot luminous cannonballs across the sky, all caused by the pieces of atoms shot at us by the Sun, accelerated toward us in association with raging solar magnetism. Far more than causing pretty lights, solar storms disrupt our planet's magnetic field, make compasses go awry, break the electronics that ride aboard billion-dollar satellites, and can even make our lights go out. When the storms subside over long periods—decades—our Earth chills. Without the Sun's magnetic effects, it's even conceivable that terrestrial life could never have begun. More invisibly, billions of subatomic particles pass harmlessly through us each second from the deeply buried nuclear furnace that powers sunlight. The Sun, Helios, indeed reaches out to touch us in ways that we are only beginning to understand.

Move on to the planets, the Earth's brethren that orbit the Sun, from Mercury close in, to Uranus and Neptune far away, the Earth number three. All are satellites of the Sun that were created along with it some four and a half billion years ago. As does the Moon, they affect us through their gravity. Too far away to cause tides of any measurable sort, they instead act to pull on the whole Earth and alter its orbit. Over the aeons, the distance between the Earth and Sun changes first in one direction, then in the other, as does the orbital oblateness. Couple that with a 26,000-year wobble in our rotation axis caused by the Moon and Sun acting on an equatorial bulge that comes from terrestrial rotation, and our seasons and climate gradually change from variances in solar heating. Ice ages may be a consequence.

Benign though they may seem, the planets even come to visit, their attentions sometimes quite unwelcome. They were born through successive collisions of smaller bodies (built from dust grains) that orbited the primitive Sun. The process, however, was far from 100 percent efficient, resulting in a large amount of leftover debris that consists of the rocky-metallic small asteroids mostly between Mars and Jupiter and the icy comets that mostly reside beyond the planetary system. Acting as "dirty snowballs," cometary bodies measured in tens or hundreds of kilometers across lie in two great reservoirs. One, a flat but thick plate beyond Neptune, is made of comets that were too thinly spread to assemble into a planet. The other, holding comets that were kicked out of the planetary system by the giant outer planets, may extend halfway to the nearest star. If caught in a long, looping orbit, a comet can get close enough to the Sun so that its ices turn to gas, which, with the release of dust, produces the graceful tails iconic to astronomy.

Collisions among asteroids coupled with the gravitational effects of planets can toss these shredded bodies to the Earth. Hitting our atmosphere and heating, they first appear as streaks of light—meteors—crossing the sky. If large enough, they smack into us as meteorites to be visited in museums. The largest can dig giant craters in the ground, and in the most devastating of such events even wipe out whole species of life. Small rocks and dust grains flaked off comets produce a steady rain of such meteors, sometimes showers of them, sometimes whole storms of them to be admired. And comets can hit us too, with awesome force. Giant impacts on other planets can be so energetic as to launch rocks into orbit around the Sun, some of which eventually also hit us, giving us a cheap "space program," whereby we can take pieces of the Moon and Mars into our laboratories. Even stars can have such effects by gravitationally disturbing the outer icy cometary bodies, tossing them back toward us.

The Sun shines the same day after day. Inside is a highly controlled fusion machine that converts the solar hydrogen fuel into helium with the creation of energy. As the fuel runs out over long periods—billions of years—the Sun will gradually change. For a time, it will swell to swallow Mercury, perhaps Venus, perhaps in the ultimate solar effect even consuming Earth. It will then quietly shrink to a cooling dense ball comparable to our own planet's size. Other stars are not so lucky. Rare massive ones explode, whereupon they accelerate their surroundings outward at speeds close to that of light. The stuff rains down upon us as "cosmic rays" that manufacture the radioactive carbon that archaeologists use to date ancient ruins. Cosmic rays may even be the seeds that trigger lightning bolts, perhaps even some cloud formation, all from dying stars that are thousands of light-years away.

If near enough, the radiation from such explosions may be hazardous to life, indeed may even have caused one or more extinction events on Earth. Ordinary stellar explosions need stay at least 30 light-years away, while ultra-rare maximum detonations could damage us from thousands of light-years away. No wonder, since their effects can even be seen with the naked eye from *billions* of light-years' distance. The remains of these destroyed stars can be just as dangerous. Stellar blowups leave behind dense remnants that pack the mass of the Sun into balls no more than a few tens of kilometers across, into neutron stars that can have magnetic fields a million billion times stronger than Earth. Adjustments in these extreme stars and in their magnetic fields send out bursts of radiation so powerful that even though tens of thousands of light-years away, they can turn off orbiting satellites and disrupt communication.

In the ultimate connection, we came from out there. All the subatomic particles that make atoms and therefore ourselves were created in the Big Bang, the event that began our Universe. All the heavier chemical elements were made in aging and exploding stars, which also produced much of the energy needed to drive star and ultimately planet formation. We are not just *in* the cosmos, we are *of* the cosmos, we, along with all the other parts of it *are* the cosmos, all of it one, all of it allowing us to be born and to live our lives so as to understand and appreciate its grand beauty.

Chapter 2

Tides of Life

A magical day at an ocean beach begins with watching the power of wind-driven waves as they roll in and crash at the shore, only to be replaced, ceaselessly, by more. Then a seemingly mysterious effect captures the eye. Ever so gradually, the sea and its waves advance on you, the water at any one point becoming deeper until it may seem that the whole beach will disappear beneath it. And then just as mysteriously, the advancing front stops, turns around, and goes backward, exposing more and more of the beach until you can walk well out into the ocean bed looking for a new load of shells.

These *tides* vary mightily from place to place, from barely noticeable to dramatically large. But wherever they are, they are remarkably regular, rising and falling day after day, month after month, year after year, successive high or low tides repeating on an average interval of 12 hours and 24 minutes, acting as rough natural clocks. Even a casual observer quickly correlates this interval with the passage of the Moon across the sky. Closer examination reveals that the tidal give and take also depends on the lunar phase, the effect lowest at the quarters, highest at new and full, and even (once we can get the measure) on lunar distance. Somehow, this companionable body that goes around us and illuminates our nights also has the ability to pull the waters, one that involves neither magic nor astrology. Tides, which have the power to control how we live, may even have helped originate our journey on Earth, one that began with the first life billions of years ago.

Isaac Newton, who lived and worked in the late 1600s, was the first to figure them out. But two items come first.

Figure 2.1. The Earth's waters, its waves and tides, rush onto the Pacific shore.
J. B. Kaler.

(1) Gravity

Isaac Newton gave us three rules through which we can understand how and why we move. First: put a ball on the ground and it
will sit there forever unless it is deliberately attacked with a "force"
of some sort. Once the ball is rolling, you have to apply another
force to get it to change its speed and/or direction—to "accelerate"
it—else off it goes in a straight line forever. Second, the degree of
a body's acceleration depends on the force you give it divided by
its mass. Third, you push on something and it pushes back. Now
apply them.

Every body in the Universe, said Newton, attracts every other
body in the Universe. Indeed, it was the acceleration (change in
direction) of the orbiting Moon around the Earth compared with
the accelerating drop of a falling body that led Newton to his great
discovery of how gravity works in the first place. Place two balls in
space. The attractive force between them that draws them together
with ever-increasing acceleration depends on their masses and on

the distance between their centers. Specifically, it depends on the product of the masses (you multiply them together) divided by the square of their separations. Cut the distance in half and the force quadruples, double it and it is cut to a quarter; increase the distance by a factor of 10, and the force is just one one-hundredth of what it was. But it can never go to zero.

As a result, the bodies fall together, accelerating with increasing speed until they bang into each other. Replace the two balls by you and the Earth. Both possess this gravitational force, gravity always mutual. Jump off a table, and you and the Earth attract *each other*. You fall to Earth, but the Earth also comes up to meet you, both moving ever faster. Of course, being vastly the more massive, Earth dominates, so in effect it seems not to move at all such that *you* do all the falling. When you stand on it, you and the Earth keep attracting each other, but the solid surface prevents you from falling further, and the force is now called your *weight*. While falling, however, there is no ground-support, and you are weightless.

In the early twentieth century, Newton's rules were redefined by Albert Einstein, who more accurately viewed gravity as a distortion in a tightly knit fabric of conjoined space and time ("spacetime") caused by the presence of mass. Other effects of this concept of "relativity" include the distortion of time itself. In most applications, Newton's much simpler rules work fine, but under extreme conditions of high mass (or high speed) one must use Einstein's. That said, no one yet knows what actually *causes* gravity, why there is a force, or why mass distorts spacetime. For all our success in developing and using the rules, we remain ignorant of gravity's origin.

Whatever its source, it's gravity that keeps the Moon in orbit. Unless a body is disturbed, it will perpetually maintain any given motion (Newton's "first"). The Moon would escape in a straight line were it not for the gravitational attraction between it and Earth that makes it want to fall toward us. But though it falls, it's also moving at high speed perpendicular to the line that connects us. The result is a curved path that is traced over and over. Though truly falling, it cannot ever reach us. The planets go around the Sun, other moons around their planets, two stars around each other, for

Figure 2.2. A lonely astronaut, untethered from his spacecraft, falls toward Earth. But since he is moving along at 29,000 kilometers (18,000 miles) per hour, the Earth curves below him at the same rate at which he drops, keeping him safely in orbit. NASA.

the same reason. A spacecraft goes around Earth because at launch it is accelerated to very high speed parallel to the ground. Cut the engines at high altitude and the craft drops, but because it is going so fast, it cannot get back home, as it is falling in concert with the curvature of the Earth below it. Since the astronauts inside are falling too, with the same acceleration as their cabin, they are weightless even though within the strong gravitational pull of our planet.

But again, gravity is mutual. Not only does the "falling" Moon go around us, we fall around the Moon, both of us circumscribing a point in between, the "center of mass." But since the Earth is 81 times more massive than the Moon, the center of mass is 1/81 the distance from the Earth to the Moon, that is, the Moon moves in a path 81 times bigger than Earth's. The Moon, averaging 384,400 kilometers (238,900 miles) from us, swings around the center of mass with an average orbital radius of 379,700 kilometers (235,900 miles), while our path is just 4700 kilometers (2900 miles) in radius. Our own motion is simply so small that it is not readily noted. The tides are a secondary effect upon the Earth of the gravitational attraction of the Moon—and of the Sun as well.

(2) Phases

Understanding the tides requires understanding the lunar phases. The big problem with the phases—changes in apparent shape—of the Moon is that the sky appears in two dimensions. The Moon seems to be at the same distance as the Sun, whereas the Sun (at 150 million kilometers, 93 million miles) is really 400 times farther. Since the Sun is also 400 times larger than the Moon, the two have the same angular diameter, half a degree. Once one visualizes the third dimension, the origin of the phases quickly clarifies.

All planets and their moons shine to the eye by reflected sunlight. Half our world is bright with daylight, while the other half, hidden from the Sun, is in darkness. The Earth rotates once a day in the counterclockwise direction (as viewed from above the north pole). We, going along for the ride from west to east, pass from day to night and back again. Since we feel stationary on our smoothly spinning planet, the Sun (and everything else) appears to move across the sky in the opposite direction, from east to west, rising above and setting below the horizon as it goes, sending daylight around the Earth. The Moon and the other bodies of the Solar System all have similar daytime and nighttime sides. The lunar phases are caused by our seeing changing proportions of lunar daylight as the Moon goes around us.

The Earth orbits the Sun in the same direction in which our planet rotates, which makes the Sun appear to move at the rate of a degree per day from west to east along its tilted "ecliptic" path, opposite the direction in which it rises and sets, and against the distant background of the ancient constellations of the Zodiac. Orbital completion takes a year, 365.24 days. The Moon orbits the Earth in the same direction, close to, but not quite on, the ecliptic plane, which also makes the Moon appear to move to the east relative to the background stars. Closer to us than we are to the Sun, its apparent motion is much faster.

Once every 29.5 days, the Moon passes between us and the Sun. Since the Moon's daylight side is now facing the Sun and turned away from us, we see only lunar night, and the "new Moon" disappears from view. The tilt between the lunar and terrestrial orbits is about 5 degrees, so usually the Moon passes under or over the Sun. If the Moon passes exactly in front of the Sun, which it does at least twice a year, someone on Earth witnesses a solar eclipse.

As the Moon moves forward in orbit, we first see a sliver of lunar daylight in the shape of a crescent. Since the Moon has not gone far from the Sun in angle, the crescent—which must face the Sun—sets shortly after sunset. And do look, as you can then see the rest of the lunar disk, most of it in night, lit by sunlight reflected from Earth. Night by night, we see more lunar daylight and the crescent grows ("waxes") fatter, while the setting time becomes later. After a bit over a week, the Moon is 90 degrees to the east of the Sun, and we see half daylight, half night, a "half Moon," but since the Moon has gone through a quarter of its orbit, the phase is also called "first quarter." And it sets a quarter of an Earth-turn after the Sun, or about midnight.

In the week to follow, we now see progressively more of the daylight side in a waxing "gibbous phase," the Moon rising in the afternoon and setting after midnight. After a half turn of 180 degrees, the Moon is opposite the Sun, and there it is, the full daylit lunar face, the full Moon, its nighttime side now directed away from us. Opposite the Sun, the full Moon rises at sunset, sets at sunrise. If the alignment is exact, the Moon passes through the Earth's shadow and is eclipsed; usually, the full Moon is just a bit north or south of the shadow. We then just see all the phases repeated backwards through the waning gibbous, to third quarter,

Figure 2.3. A crescent Moon, setting in evening twilight and visited here by Venus, shows off a sliver of lunar daytime. J. B. Kaler.

which rises around midnight, is seen in the west during the morning, and sets around noon. The show ends with the waning crescent, which—rising ever later after midnight—gradually disappears into dawn. Finally, 29.5 days later, after a "lunation," the Moon arrives back to its new phase and the cycle—which is the origin of the "month"—starts anew.

Tidal Stretching

The gravitational force between two bodies depends on their distances from each other, specifically that between their centers. But different parts of the Earth are at different distances from the Moon, and are therefore subject to different strengths of lunar gravity. The force is strongest at the "sublunar point," where the Moon would appear directly overhead, and weakest exactly opposite, on the other side of the Earth. Hard as rock may seem to anyone falling on the ground, it's still malleable, especially within the high-pressure, heated terrestrial interior, where it softens. The difference in gravity across the Earth's body thus causes our planet to be stretched rather like taffy on a line toward the Moon, that is, the Earth—the *solid* Earth—will have a small "tidal bulge" of about 20 centimeters (8 inches). As the Moon reaches toward its highest point in the sky, you rise up a bit and then settle back down. The effect is not in the least personally noticeable (nor has it any effect on you), but sensitive instruments long ago detected it.

While the solid tide is subtle, the effect on the Earth's waters is not, since they can easily flow in accommodation to the differing gravitational forces offered by the Moon. On the side of the Earth that faces the Moon, oceanic water flows toward the sublunar point. On the side of the Earth facing *away* from the Moon, it flows in the opposite direction, toward the most distant point on Earth that is exactly opposite the Moon. That is, the ocean stretches out such that at both locations, on the side of Earth facing the Moon and on the side pointing away from it, the ocean moves to deepen.

Since the water is pulled away from the direction that lies *perpendicular* to the line to the Moon, there the ocean becomes slightly shallower. Once again, the result is a tidal bulge, but now in the Earth's seas, one that at maximum is about a meter high. As the Earth rotates, you, on the beach, will pass through deeper, then shallower, then deeper water. And the ocean seems to go up and down, pulling forward to *high tide*, then drawing back to *low tide*, in a perpetual cycle. In this picture you will see a high tide with the Moon in the sky, and then once again when it is below the horizon and visible on the other side of the Earth.

You will, however, not necessarily see a meter-high tide. Shoreline effects that include the slope of the ocean bottom, funneling by landforms, and the like, make each coast, each beach, unique. At some you notice minimal effects, while at others the tide sweeps in and roars back out with great depth. In Nova Scotia's Bay of Fundy, the tide goes up and down by a full 15 meters (50 feet), while in open-ocean Hawaii, it is not much to look at. In some places, notably off the north coast of France, the tide goes out for such a distance, and comes back in so fast, that people looking for treasures have been trapped and drowned by the rushing waters.

Figure 2.4. High tide at Cutler, Maine, at left contrasts dramatically with low tide at right. Maine Department of Sea and Shore Fisheries.

Tides can be seen only in the largest bodies of water. They are invisible in lakes and inland seas, their smallness hidden by wave motion and sloshing caused by winds. The exception is in rivers that connect to the sea, where tides can be found a long way from the river mouth. The Hudson River experiences tides at Albany, well over 100 miles from the river's outlet in New York. At the extreme, an incoming tide can send a wall of water upstream in a strong wave called a "tidal bore." In some rivers bores can be dangerous, while in others they are a source of play as surfers can ride the wave for miles upstream only to float back down to await the next one. (Do not confuse any of these phenomena with the "tidal wave," a misnomer for the dreaded *tsunami*, which is caused by undersea earthquakes, landslides, or volcanoes.)

The daily tide was a powerful motivating factor in the early belief in astrology (and for that matter, still is), that the heavenly bodies affect affairs on Earth, which they do, but in ways no astrologer could ever have dreamed.

Tidal Lag

Explanations in science often go by successive approximation. The Earth is a sphere. But only if we do not look too closely. Then we see that terrestrial rotation throws the planet slightly outward at the equator, making the Earth really into an "oblate spheroid" that is 40 kilometers (25 miles) wider across the equator than it is through the poles. Look more closely yet, and even the oblate spheroid becomes an approximation to the "geoid," a complex shape that is not amenable to a mathematical formula and must be carefully mapped.

So it is with the tides. The simple explanation has the tidal bulge facing the Moon. Water, however, does not flow infinitely fast. Even if you tip over a cup of coffee, you might, if you're quick enough, turn it upright before the liquid makes too much of a mess on your desk. So while the Earth's waters do try to flow toward the line between Earth and Moon, they are still tied to the terrestrial body through friction and affected by their own viscosity, so they cannot quite get there. As the Earth rotates, it therefore drags the tidal bulge forward, ahead of the Moon. The waters try to run against the rotation, but never make it to the expected point. The result is that high tide, instead of occurring with the Moon more or less overhead, or at least as high as it gets, takes place after the Moon has passed its peak and while it's descending the western sky, the exact time depending on the lay of the land.

Tidal Clock

The rotation period of a spinning body depends on what one uses as a reference. Stand in the middle of a room. Turn to your left, and things in the room seem to move to the right. When you see Grandma's picture swing by the second time, you have completed a full rotation relative to the room's walls. Now stand outside. Draw an imaginary circle in the sky from the north point on the horizon through the point overhead (the "zenith") to the south point, tracing out a "celestial meridian" that cuts the sky in two and that, like Grandma's photo, serves as a reference. The "solar day" is the time it takes the Sun to make two successive passages across the celestial meridian at noon.

But we need not use the Sun to assess the rotation period. We might instead use the stars. Because the Sun (like the Moon) continuously moves eastward against the stellar background, it takes less time for a star to appear to make a full turn (to cross and then return to the meridian) than it does the Sun, giving us a "sidereal" (star) day that is four minutes shorter than the solar day. The sidereal day is the "true" day, the one that would be observed from the outside, say by a Martian.

Now use the Moon! The effect is opposite that of using the stars. As the Earth rotates, the Sun appears to move through an angle of 360 degrees in 24 hours, so in one hour it moves through 15 degrees. Since the Moon's orbital period relative to the Sun (the period of the phases) is 29.53 days, the Moon pulls ahead of the Sun by an average of 12.19 degrees per day, which means that the Moon crosses the meridian 12.19/15 (0.81) hours, 48.8 minutes, *later* each day than it did the day before. You will then (all things being equal) see two high tides every 24 hours and 48.8 minutes, a "tidal day," each one separated by 12 hours 24.4 minutes, with low tides in between them. Once you know when a high or low tide occurs for any given location, you can in principle predict the next one and all the succeeding ones thereafter. You can even buy a "tide clock" that will do it for you. Though there are very significant variations in the times of high and low tides that are caused by the tilt and shape of the lunar orbit, the form of shoreline, and the effect of the Sun, over the months, they nicely average out.

Now Add the Sun . . .

The Moon is small compared with Earth, only a quarter its size and 1/80th its mass. It has a powerful tidal effect on us because it is so nearby, the closest permanent body to us. Asteroids and comets can get much closer, but their proximity is temporary: they either pass us and zip back out to distant space, or they hit us. While a strike can wreak havoc, these visitors are far too small to produce significant tides. We will encounter them later.

While the Sun is 400 times farther from us than the Moon, it is so massive (333,000 times more so than Earth, 2.7 million times more than that of the Moon) and exerts such a powerful gravitational attraction that it produces strong terrestrial tides as well, though the great distance reduces them to about 45 percent that of the lunar tide. The Moon thus contributes two-thirds of the total, the Sun the remaining one-third.

The solar tide is completely independent of the lunar tide. Neither, in a sense, knows anything about the other. When the Moon and Sun are lined up, at *either* new Moon or full Moon, the lunar and solar tides add together, which maximizes the effect at the world's shores, yielding both the greatest high tides and the greatest lows of the lunar month, or phase cycle. Though called "spring tides," such maximized tides have nothing to do with the seasons, and occur twice every 29.5 days.

As the Moon orbits and waxes into crescent from new (or into gibbous from full), the tides weaken as the solar tide begins to fill in the lunar tide. At the quarters, the two tides work counter to each other, and the tides—known as "neap tides"—are minimized, dropping to about a third those of the spring tides. One cycle, that of the phases, then gets piled on top of another, that of the Earth's rotation as coupled to lunar orbit. The effect has been noticed for as long as humans have walked the Earth.

. . . And Elliptical Orbits

An ellipse is an oval with a precise definition. Place two nails in a board and attach a loose string to each. Stretch the string taut with a pencil. Keeping it tight against the string, start to draw, and you get an ellipse, wherein each nail is a "focus" of the figure. In 1610, Johannes Kepler made one of the great scientific discoveries of all time, that planetary orbits around the Sun are not circles but ellipses, with the Sun at one focus (Kepler's first law), the other focus meaningless. Merging the two nails into one yields a circle, which in nature is impossible to achieve exactly (though we can come very close). In a circular orbit, the distance between bodies is constant, but in an elliptical orbit, the separation continuously changes.

While the Earth's path does not deviate much from circularity, just 1.7 percent, the effect is readily measurable. At "perihelion," which takes place around January 2, we are 1.7 percent closer to the Sun than average, and about July 3, when we pass "aphelion," we are 1.7 percent farther away, for a total swing of 3.4 percent. (These dates have nothing to do with the seasons, which we will encounter in the next chapter.) The Moon's orbit around the Earth is notably more eccentric, 5.5 percent. At "perigee," which the Moon passes about once a month, the Moon is some 11 percent closer to us than it is at "apogee," where it is farthest. Since there is no point of reference, the effect cannot be seen with the naked eye, though it is easily photographed. (Distance variation is not coupled to phase, and has nothing to do with the Moon seeming to loom large when on the horizon: that is strictly an optical illusion.)

Distance profoundly affects tides. While the gravitational force behaves according to the inverse of the square of the distance between two bodies, a tide—a differential effect between the gravity on one side of a body as opposed to the other side—behaves according to the inverse of the *cube* of the separation. Cut the distance in half and the tidal effect goes up by a factor of eight. As a result, since the distance between us and the Moon changes by 11 percent, a factor of 1.11, the tide at perigee is 1.11 × 1.11 × 1.11 = 1.37 times, a whopping 37 percent, stronger than it is at apogee!

We therefore pile a third cycle atop the first two, one now related directly to the lunar orbit. When full or new Moon takes place near perigee, the amplitudes (the differences between high and low tides) of the spring tides are notably strengthened, giving us maximum monthly highs, while if either takes place at apogee, they are markedly weakened.

Finally, return to the Sun. While the distance variation is not as large as it is for the Moon, it's still significant. Forgetting the Moon for the moment, spring tides near perihelion in December and January are 10 percent stronger than they are near aphelion in June and July. The greatest of all spring tides will take place when the Moon goes through perigee in early January or late December. The weakest of all will be neap tides at lunar apogee in late June or early July.

The tidal flow is affected not only by shorelines but also by how high the Moon is in the sky at meridian passage, which in turn depends on where the Moon is in its orbit and by latitude (angular distance from the equator). Many are the vagaries and surprises. There are locations, for example, that witness tides at lunar perigee, but not at apogee.

Tides are fun to watch, the interacting cycles quite obvious over the year to a shore dweller. In the days of sailing ships, they were greatly important, as ocean craft would naturally float out to sea with the receding tide. Unless you are tempted to walk too far out in the ebbed tide at places where it comes back in with a roar, or onto a sand bar where you might get cut off from dry land, tides in themselves are not dangerous. However, winds off shore can pile up coastal water through wave action. Coupling of a high tide with a storm, worse a hurricane, can be disastrous. The absolute worst is a storm that coincides with full or new Moon when perigee also takes place in early January, which can result in heavy flooding and damage. It's happened more than once.

"Time and Tide . . ."

. . . begins the cliche. The two, however, are deeply rooted in each other, though in ways quite different from the original meaning of the phrase, or for that matter, from the tide clock. Tides in fact give fits to those trying to tell time.

Time. Our time of day begins with the position of the Sun in the sky. Start the day at 12 noon with the Sun on, then moving past, the celestial meridian. For every 15 degrees of Earth rotation, the Sun moves 15 degrees farther to the west of the meridian and we add another hour to the clock, each hour divided into 60 minutes, each minute into 60 seconds. After half a terrestrial rotation, the Sun has moved through 180 degrees, the time has advanced by 12 hours, the clock reads midnight, is reset to 0 hours, and we start the count all over again.

Such time is told by a shadow-stick, a sundial. The eccentricity of the Earth's orbit and the tilt of the rotational axis, however, make sundial time variable: as the year progresses, the hours have different absolute lengths. After correcting for these effects, adding heavy layers of technology that actually have us observing distant stars and galaxies instead of the Sun itself, we arrive at a smoothed-out "mean solar time."

Mean solar time depends on where you are on Earth. If you move directly north or south (along a meridian of longitude), the Sun will respectively move south or north, but will keep the same position relative to the meridian. People north or south of each other always keep the same mean solar time. But if you move directly east (along a parallel of latitude), the Sun appears to move west relative to the meridian, and the time gets later; if you move west, the Sun conversely appears to move east, and the time gets earlier. Which is why it is earlier in California than it is in New York.

Raw mean solar time is difficult to use because it changes continuously with east-west location. At a latitude of 45 degrees north (halfway from the Earth's equator to the north pole), someone standing a kilometer (0.62 miles) to the west of you has a clock that reads 1.6 seconds earlier than yours, another person 100 kilometers distant, a clock over two minutes earlier. To give consistency to life and airline schedules, we artificially divide the world into 24 time zones, each in principle 15 degrees wide, within which everyone happily agrees to keep the time of the central position (though in practice the time zone boundaries are greatly distorted for social and political reasons). As you move from one time zone to the next, this brilliantly conceived "standard time" (actually invented to accommodate railroad timetables) changes by an hour, but the minute stays the same. The worldwide standard is the time at the prime meridian (zero degrees longitude) that runs through Greenwich, England, which goes by Greenwich Mean Time (GMT), or more commonly, Universal Time (UT)—a bit arrogant, since it is doubtful it is used by residents of Arcturus. Ninety degrees to the west of Greenwich, the Standard Time is 6 hours earlier than Greenwich, 90 degrees to the east, 6 hours later.

Enter the Moon. While the Earth's rotation drags the tidal bulge ahead of the Moon, the effect does not stop there. From the Moon's point of view, the Earth is not symmetrical, but has a tidal bulge against which lunar gravity constantly pulls. The Earth's waters, the bulge included, are tied to the ocean beds and the solid Earth through a sort of "friction" that includes the interaction with the shorelines. Acting as a celestial brake, lunar and solar gravity therefore drag back on the Earth's rotation, which is slowed down. And the day becomes longer.

The effect was first noted by comparing the predicted times of ancient eclipses with those actually witnessed and recorded. Modern observations (of stars and distant galaxies) show that the day lengthens by 0.0014 seconds per century. But we do not add seconds to the solar day, which always contains 24 hours, 1440 minutes, 86,400 seconds. Instead, the lengths of solar units slowly increase. While the effect may seem negligible, it has a profound effect on a running clock, which continuously accumulates the error, allowing us to note the change from one day to the next. Piled on top of the general slowdown are a variety of erratic effects that produce both rotational decelerations and accelerations and that include winds shoving on mountains, earthquakes, and changes in the Earth's internal structure. How then are we to keep a high level of science and technology in a modern world if the terrestrial time standard is continuously changing?

First, define the absolute length of the day, hour, minute, second by those for a specific year to give the second a definition of constant length. Then divorce time units from the Earth's rotation altogether, and define them in terms of unchanging atomic vibrations. The "atomic clock" runs smoothly with no erratic or systematic variations. But we live on the Earth, and clocks are supposed to tell us where our planet stands in its rotation. If we were to run entirely on the atomic clock, the times of sunrises and sunsets would gradually get out of synchrony with each other. Moreover, without clocks tied to the Earth, astronomers could not accurately locate heavenly bodies. So we use a hybrid called "Coordinated Universal Time." During daily life we run on the atomic clock, but keep our eye on the sky. When the Universal Time (as told by the apparent positions of stars and galaxies) gets up to 0.9 seconds behind Atomic Time, we add an extra second to the clock, a "leap second," that brings the two back into harmony, the leap second a direct descendent of the perpetual tides. Leap seconds are usually added at the end of the year, when one particular minute will contain 61 seconds rather than 60, compliments of the Moon and Sun.

How High the Moon?

As the Earth in its elliptical orbit gets closer to the Sun for its January perihelion, it speeds up; as it approaches its July aphelion, it slows (Kepler's *second* law). From a calendar, count the days of the seasons. Northern-hemisphere winter falls between December 21, when the Sun is as far south as it can get, and March 20, when it crosses the equator; northern summer lies between June 21, with the Sun as far north as it can get, and September 23, when it again crosses the equator. Though it hardly seems so, as a result of the orbital eccentricity and the accompanying velocity change, northern winter is actually a few days shorter than northern summer.

The speed variation is explained through the concept of "angular momentum." Tie a rock to a string and whirl it in an "orbit" about your head. The rock's angular momentum is the product of its mass, its velocity, and the string's length. In a closed system, one with no outside influences, angular momentum is "conserved," that is, constant. Reduce any of the three values that enter into it, and another must increase to compensate (and vice versa). As the distance between a planet and the Sun decreases, the orbital velocity goes up, and Kepler's second law is explained. (Such variations in the Moon's orbital speed help confound the tide clock.)

The angular momentum of a rotating body is the sum of the angular momenta of all its smaller parts as they circle about the rotation axis. The tightly bound Earth-Moon system is a decent approximation to a closed system within which the total rotational and orbital angular momentum must stay the same. Slow the Earth through tidal friction, and its angular momentum decreases. In response, the Moon's orbital angular momentum *in*creases, which it does through an increase in the distance between it and the Earth.

The distance to the Moon can be measured precisely by bouncing laser beams off mirrors left on the lunar surface by Apollo astronauts and Soviet landers: it's just the one-way flight time divided by the speed of light. The current outbound movement is found to be 3.8 centimeters (1.5 inches) per year, which in turn ever so slowly increases the period of the lunar orbit and the interval between the tides. In the distant past, the Moon must have been considerably closer to us than it is today and the orbital period and tidal interval shorter. The period of the phases, 29.5 days, is the orbital period of the Moon relative to the Sun. Since the Sun is continually moving east relative to the stars, it takes the Moon less time to cycle around relative to a star than it does relative to the Sun. The "sidereal period," the true orbital period, the one that would be seen by our favorite Martian, is only 27.3 days. A study of fossilized tidal shorelines shows that 3.2 billion years ago, when the world was only 1.4 billion years old (the Earth, from analysis of radioactive chemical elements in Solar System rocks, 4.6 billion years of age), the sidereal period was only 20 days. Orbital theory gives a lunar distance 80 percent of what it is today, and within a factor of 2 fits with the current measured rate of change!

Imagine yourself in those ancient times when life was just getting under way. The Moon, glowing in the primitive sky 25 percent larger than you see it today, zipped through its phase period in only 21.1 days, yielding 17 lunar months per year rather than the current 12.4. Since the strength of a tide depends on the inverse of the cube of the distance between the bodies, these ancient lunar tides were about double the size they are now. According to some theories, the resulting heavy "tidal stirring" was quite possibly a strong factor in the initial creation and development of life itself!

We live in a special time in which the Moon's angular diameter of one-half of a degree is almost exactly that of the Sun's (ignoring slight changes caused by distance variations that in turn are caused by elliptical orbits). As a result, about twice a year, the Moon crosses exactly in front of the solar disk to give us a solar eclipse that allows the study of the outer solar layers. As the Moon moves away from us, its angular diameter is becoming smaller. About 100 million years from now, at its very greatest it will have shrunk to that of the Sun, after which total solar eclipses will be impossible. So enjoy them now!

It's Mutual

Gravity is always mutual, and so also must be the tides. As the Moon raises tides in the Earth, the Earth has to raise them in the Moon. There is no lunar water, no oceans to slosh around (no water at all in fact except maybe for some inconsequential ice at the poles), so tides in or on the Moon must be restricted to the solid body. The size of a tide responds to the difference in the gravitational force across one body as produced by another. The Moon is only a quarter the size of Earth, so all things being equal, the tide raised by the Earth on the Moon would be smaller. But all things are not equal, as the Earth has 81 times the mass of the Moon, resulting in much higher tides on the Moon than we could ever have here.

Whether there is water or not, the constant tidal flexing of a body dissipates rotational energy and slows the spin rate. The Moon and Earth have been together practically since the birth of the Solar System, the Moon most likely forming in a giant collision between the primitive Earth and a competing Earth-crossing planet. We won. The debris became the Moon. However fast the Moon may originally have been rotating, tides raised in it by the Earth very quickly slowed it to the point where it kept one face pointed toward us at all times, as it does today, the tidal bulge facing us, the lunar rotation locked onto us.

Look at the dark regions, the spots that make the "Man in the Moon" and other fanciful figures. They are ancient lava-filled impact basins and flows akin to the much smaller craters seen through telescopes (and that were made by an early heavy bombardment from comets and asteroids). With no erosion from wind (no atmosphere), water, or ice, unless battered by further impacts they will remain untouched practically forever, whereas Earth's craters were, and still are, quickly wiped away. And that is just the way we see the lunar features in the sky, forever constant in position, as the Moon keeps one side toward us (the "near side"), the other always pointed away from us (the "far side"). (In truth, because of the tilt and eccentricity of the lunar orbit, we see about 60 percent, only 40 percent really completely hidden.) Such synchrony is near-universal

Figure 2.5. Coupled forever, a cloudy, life-filled Earth and its battered, barren Moon mutually orbit a common point between them. Tides force the Moon to keep one face toward Earth, while the Moon slowly returns the favor by slowly retarding Earth's rotation. *Apollo 8*, NASA.

among the many moons of the Solar System. Since the lunar far side is perpetually invisible from Earth, we had no idea of what was there until we could send spacecraft around to see it. No surprise, it looks rather like the near side, except for the lack of the large dark spots, a difference that is still not well explained.

The Moon does, however, rotate with respect to the Sun, the Moon's "day" that of the period of the phases, giving any location a sunrise, a daylight period of two weeks, a sunset, and an equally long night. During the waxing lunar phases, you see the sunrise line cross the lunar disk as it goes from crescent to first quarter to gibbous to full. Through the telescope you can see mountain peaks (really crater walls) catching the first rays of the warming Sun. After full, during the waning phases, you see the sunset line bringing lunar night.

If you lived on the near side of the Moon, you would see the Earth hanging in your sky, unchanging in position, but going through phases just as the Moon does from Earth, though half a turn different, "new Earth" taking place at full Moon, "first quarter Earth" at third quarter Moon, and so on. If you were a native "lunarian," one who grew up on the far side, you would be completely unaware of Earth. The only sign of it would be that your astronomers would detect apparent shifts in the positions of the planets and the Sun, showing that the Moon was orbiting about something large. Then you would send out your explorers, who finally cross over to the near side to see the glorious blue-white ball looming up above the horizon. Maybe someday someone really will live there. Out of sight of Earth, the far side provides a dark place from which to observe the sky and a radio-quiet one that would delight radio astronomers.

As Earth slows and the day lengthens, it will try to synchronize its rotation onto the Moon, as the Moon's already has onto the Earth. Full synchronization has already taken place between Pluto and its large satellite Charon, the two hanging in each other's skies for the rest of time. We will never see that happen, however, because the time scale for our Earth-Moon system is longer than the stable age of the Sun, which as it dies will expand outward to meet us and make a mess of the whole situation (as we'll see in chapter 7).

But that is in the far-distant future. For now, the Sun touches us not just through the tides, but also through its internal nuclear reactions and the effects of its powerful magnetic fields. For now, uncoupling the Sun from the Moon, we next see what else it can do both for us and to us. The Moon will come back to us later.

 Chapter 3

Solar Storm

Why is the Earth round? Gravity. Gravity draws objects together, makes things fall to Earth, makes orbits possible, creates the tides. But gravity works its magic on single bodies too. All the parts and particles within the Earth pull on all the other parts and particles, and not only keep the Earth together but draw it into a near sphere. Raise any portion of it upward, and self-gravity will snap it back, as it will on all the other planets as well. Gravity enforces sphericity.

There are exceptions that first include the Earth's rotationally induced equatorial bulge. The outward force is too great for gravity to bring the bulge back inward, and the two effects simply achieve a balance. Second, rock is strong. A pile of it can nicely resist gravity's inward pull. The result is our planet's beautiful and challenging mountains. But there is a limit. Above about 10 kilometers (7 miles), the base of the mountain begins to collapse under the vast weight on top. No bigger mountain—from base to top—can exist. The "Big Island" of Hawaii is the ultimate result. Only when bodies are under a few hundred kilometers in diameter can the strength of rock win out, allowing small comets, asteroids, and moons to take on irregular shapes.

Within the Earth, each layer must support the weight of all the other layers above it, and the interior thereby compresses to higher density. Temperature also increases with depth. While the Earth's surface crust is cold, deep diamond mines are killingly hot. Kilometers down, the rock becomes so hot that it turns plastic. Relieve the pressure and the rock liquifies and explodes through tubes and cracks in the crust in volcanoes, sure evidence of the immense heat inside.

The Earth was created and first heated just over 4.5 billion years ago through the violent collisions of countless smaller bodies that orbited in a disk surrounding the primitive forming Sun. For a time the Earth became a giant liquid, or at least plastic, blast furnace that sent heavy iron, nickel, and the materials that dissolve in them, to the inner-Earth, where it resides today in a mostly iron core half our planet's size, the heat maintained largely by the decay of radioactive elements. Atop the core is lighter rock, of magnesium, aluminum, calcium, silicon, iron, oxygen, and other chemical elements that at the surface create our myriad minerals. So great is the internal heat—over 7000 degrees Celsius, hotter than the solar surface—that the outer half of the iron core is molten, the huge pressure within the inner half compressing it back to a solid.

Distinct from Earth, the Sun, 150 million kilometers (93 million miles) away, is a gas throughout. The apparently "solid" surface is just an ultra-opaque gas. Unlike Earth, the Sun (as are nearly all other stars) is made almost entirely of the lightest of chemical elements, hydrogen (by number of atoms, 92 percent at the surface) and helium. The heavy stuff out of which the Earth is made constitutes a mere 0.15 percent (the remaining 8 percent helium). With no rock to be piled up, and a slow 25-day rotation, the Sun squeezes itself into a near-perfect sphere 150 million kilometers (860,000 miles) across, 109 times the width of Earth. As a result of its great mass, 330,000 times Earth's, it also compresses to a whopping central density eleven times that of lead and a temperature of 15.6 million degrees Celsius that keeps all solar matter in the gaseous state.

The conditions are so extreme that within the inner quarter of its radius, the Sun acts like a giant nuclear fusion machine that turns hydrogen into helium with the creation of energy that slowly leaks to the 5500°C solar surface and is radiated away as the heat and light that warm us and make all life possible. Along the way from the central engine to the solar surface, the Sun develops a powerful magnetic personality that affects us, sometimes subtly, other times with the force of a celestial sledge hammer. But so do the internal reactions affect us, and not just through the heat and light that give us life.

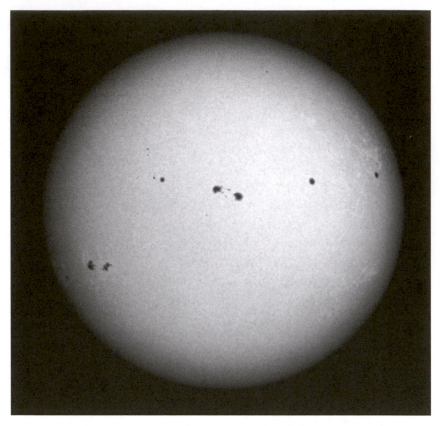

Figure 3.1. Magnetic sunspots splay across a "white light" (as opposed to filtered) image of the Sun, a ball of hot gas 1.5 million kilometers across. The larger spots could absorb the entire Earth. The dimming near the edge is the result of the partial transparency of the apparent solar "surface," within which the temperature rises with depth. Far below, in the inner core, hydrogen being converted into helium provides solar energy and keeps the Sun stable, as it will for billions of years into the future. NOAO/AURA/NSF/WIYN.

Neutrinos Going to Get You Even If You *Do* Watch Out

But you won't feel a thing.

Gravity is but one force of nature. Only three others are known. Of them, the most familiar is the electromagnetic force. Unlike gravity, electric charge comes in two flavors, positive and negative, which attract each other. Atoms are built of tiny particles.

At their cores are balls of positively charged protons and neutral (no electric charge) neutrons. Unless they are ripped away by high temperatures, these "nuclei" are surrounded by rapidly orbiting negative electrons that are attracted to the positive protons and that make the atoms as a whole electrically neutral. The different chemical elements have different numbers of protons in their nuclei: hydrogen 1, helium 2, carbon 6, iron 26, uranium 92. Nuclei are inconceivably small: you contain around 50,000 trillion trillion protons and neutrons. Different elements also have different numbers of neutrons, hydrogen usually none, helium usually two (giving its nucleus four particles), iron usually 28. Their number is not entirely fixed, however, giving each element a set of different "isotopes." Hydrogen, for example, has an isotoptic variant with an attached neutron called "deuterium."

The particles in an atomic nucleus are bound together through the most powerful force of nature, the short-range "strong force" (which keeps the naturally repellant positive protons from flying away from each other). If there are too many or too few neutrons in a nucleus, it turns unstable and begins to fall apart with the release of energy in the form of dangerous gamma rays (very high-energy light) and ejected particles, that is, it becomes "radioactive." The steady decay of uranium and thorium (90 protons) into lead helps heat the Earth and provides a natural clock with which to date its rocks, Moon rocks, and meteorites (small asteroids that hit Earth).

Helium is not quite four times as heavy as hydrogen, the deficit just 0.7 percent. Under conditions of extreme temperature and density, four protons can be forced to overcome their mutual antipathy and be merged together under the strong force to become helium, two of the protons becoming neutrons in the process. The lost mass emerges as energy. Remember $E = Mc^2$? E is energy, M is mass, and c is the speed of light, a huge number, 300,000 kilometers (186,000 miles) per second. A small amount of mass thereby becomes a large amount of energy in the form of high-energy gamma rays. Nearly all solar energy comes from this source. Given the luminosity of the Sun—400 trillion trillion watts—the process must turn 600 million metric tons of hydrogen fuel into helium ash. It has been doing so for 4.6 billion years, and there is enough core hydrogen left to make it go for another 5 billion. All stars work this way, or did before they died, or if they are forming, eventually will.

The main solar fusion process requires three steps:

- two protons to make deuterium
- the collision between a proton and a deuterium nucleus to make light helium (with but one neutron instead of two)
- the final mating of two light heliums into a normal helium with the ejection of two protons that are free to participate in the process all over again

To complete the first step, to make the deuterium, one of the protons must turn into a neutron. It does so through the fourth of our forces, the "weak force," by giving up its electric charge through ejection of a positive electron (a "positron"), a form of "antimatter" that whacks a normal electron to produce more gamma rays. The conversion also releases an even tinier particle that has almost no mass at all, a "neutrino," Italian for "little neutral one."

Energy is carried by light in all its forms (from ultra-high-energy gamma rays through X-rays, ultraviolet, ordinary lower-energy optical light, low-energy infrared, to ultra-low radio). But it is also carried by neutrinos. Light runs at the speed limit of the Universe; nothing is faster. Einstein showed that only massless particles can go at the speed of light, c. Neutrinos, while having some mass, can't get there. But they are so *very* light that they can go *almost* at c, so close that the difference has never been measured in the laboratory and has to be inferred from natural events.

The energy created in the solar interior is trapped by the high-density, high-temperature gas, and only slowly works its way out as it is cyclically absorbed and re-emitted at successively lower energies by the atoms that stand in its way. Today's benign sunlight was made in the nuclear furnace nearly a million years ago. Neutrinos, though, being weak-force particles, only barely interact with matter. Not bothered by the Sun's great overlying mass, they escape immediately in a straight line. Eight minutes after being formed, moving achingly close to the speed of light, they are here at the Earth. At the known rate of fusion, that required to produce the solar luminosity, the Sun creates 100 trillion trillion tril-

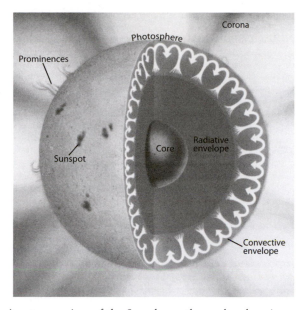

Figure 3.2. A cutaway view of the Sun shows the nuclear burning core where
the conversion of hydrogen into helium creates the solar energy and, as a
by-product, a flood of neutrinos. The "radiative envelope" transmits the solar
radiation, then hands it over to the convective envelope, which releases it into
space from the surface, or "photosphere," nearly a million years later. Neutrinos
escape immediately. Surrounding the whole affair is a magnetically heated
corona in which hang cool bedsheet-like "prominences." From *Stars*, Scientific
American Library, New York: Freeman, 1992, © J. B. Kaler.

lion neutrinos per second. As you sit reading, in that brief moment
somewhere around 100 trillion of them pass through you.

But that is the heart of the matter. With the lightest possible
of touches, they *pass through you* without stopping, without encountering any of your atoms. They pass through the whole *Earth*
without stopping. You are bathed in as many neutrinos at night,
with the Sun shining on the other side of the Earth, as you are during the daytime!

> . . . The Earth is just a silly ball
> To them, through which they simply pass
> Like dustmaids down a drafty hall . . .
> (John Updike, "Cosmic Gall")

If they pass through you, they cause no effects. However, once in a great while, a neutrino *can* score a direct hit on a terrestrial atomic nucleus (changing it into a detectable radioactive isotope of another atom) or on an electron in a water bath (causing a flash of light). Hence we can build neutrino "telescopes" to detect them and thereby look directly into the Sun's core, impossible with light, as it leaves the Sun to come to us only from the opaque solar surface. Given their known propensity for being captured and for changing their forms, they arrive in the predicted number. Day or night, it does not matter. Solar fusion really does work as theory expects. And all from solar particles that reach out to us, though the odds of *you* actually getting and keeping one (with no harm done) are perhaps but once in a lifetime.

A far more personal solar outreach, one with deep implications for humanity, involves two phenomena.

(1) Oatmeal

Nobody with any sense stares at the Sun, either with the naked eye or through a telescope, without appropriate protection. The solar brightness is so great that the naked eye would quickly be destroyed, leading to permanent blindness. While good filters are available, home-made ones (including so-called blackened film negatives) are a terrible idea. As a result, the Sun is not considered much of a "tourist attraction," and is thought of as only a brilliant, featureless, yellow-white disk.

But bring a powerful, properly filtered telescope to bear on the Sun during a quiet day with no disturbing turbulence in the Earth's atmosphere, and a detailed "sunscape" begins to reveal itself. The first thing you notice will most likely be dark spots, "sunspots," some of which may be bigger than Earth and that tend to gang together into groups, some so large they can be seen (through a proper filter) even with*out* a telescope. Once thought to be "holes" in the bright solar atmosphere through which we see down to a cool surface, they are actually vast magnetic "storm systems."

Look past the spots, and you now note that the Sun's apparent disk is not uniformly bright, but is darker at the edge (the "limb") than it is at the center. The causes are four: (1) the Sun is really a three-dimensional sphere that projects onto the sky as a 2-D disk; (2) while the gases that make the solar "surface" are very opaque, your vision still penetrates inward to a depth of a thousand kilometers or so; (3) the temperature rises inward; (4) the hotter the gas, the more energy it radiates. You look straight inward at the apparent disk's center, whereas at the limb you look at an angle and therefore not as deeply. The temperature of the observed gases through which your vision peers is less at the limb than at the center, and therefore the brightness is lowered.

Examine the Sun next under the high magnification. The surface no longer appears smooth, but becomes grainy as the solar disk breaks into hundreds of thousands of bright hotter cells surrounded by darkened cooler lanes. Though small compared with the whole Sun, the individual granules are still huge, the sizes of large U.S. states. Unlike Texas or Montana, however, they are not permanent features of the sunscape, but come and go over intervals of but a few minutes.

Their natures are found from analysis of their light. Spread sunlight into an artificial rainbow, its "spectrum" of colors from red at the long-wavelength end to violet at the short. If you have stretched the spectrum enough such that a huge array of color-shades is clearly defined, you will see thousands of fine gaps, dark "lines" ("absorption lines" because they absorb light) that look like a celestial barcode, some dark and broad, others only barely visible. Each of these lines is produced by a specific chemical element that absorbs sunlight at very specific wavelengths. Each element, indeed each ion of each element (an "ion" is a charged atom with one or more electrons removed), produces a unique pattern. Hydrogen has four lines visible to the eye, while complex atoms such as iron have tens of thousands. But each pure elemental or ionic spectrum is different and identifiable. From the strengths of these lines plus a goodly application of atomic theory, we can find temperatures, densities, and chemical compositions.

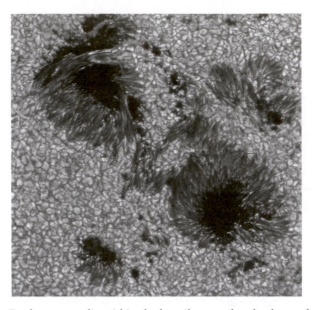

Figure 3.3. Dark sunspots lie within the heavily granulated solar surface, the granules the fleeting tops of upward-moving convection cells. Each central, depressed sunspot core is encased in a striated grayish area that connects it to the surrounding solar plain. Powerful magnetic fields that inhibit convection refrigerate the spots by some 1500 degrees. G. Scharmer and K. Longhans, Institute for Solar Physics, Royal Swedish Academy of Sciences.

The lines also let you find celestial speeds. As a radiating body comes toward you, its light waves hit you more frequently, and the colors become slightly bluer; if it recedes, you see the reverse. The greater the speed, the greater this so-called Doppler shift. Under ordinary conditions, from streetlights to stars, the color shift is far below a human's ability to detect. The absorption lines, however, mark very precise wavelengths within the spectrum. They too shift in color—in perceived wavelength—with the speed of approach or recession, and from the small degree of displacement compared with wavelengths in the lab, we can measure very exact velocities.

Examine the lines in the bright granules, and you find that they are coming at you with speeds of a kilometer or so per second, while the dark interstitial lanes are falling. The solar surface is in a state of "convection," in which hot gases rise and cooler ones fall, not unlike the air in a closed, heated room. The Sun is boiling like a pot of oatmeal on a stove.

Similar analysis of up-and-down solar motions reveals sound waves that make the Sun appear to ring like a bell with many overtones. Theory applied to the solar ringing indicates that the convection layer is hardly superficial, but extends a third of the way into the solar sphere. Convection is the primary way in which heat is brought through the outer solar layer to the surface, whereupon it is radiated away.

(2) Spin

While sunspots are not permanent features of the Sun's broad vista, they last long enough for us to see them parade across the solar surface. They are not moving through the solar gases, but *with* them as the Sun rotates on its axis something like the Earth, but with a much longer period (at its equator) of 25 days. Doppler measures are consistent in showing an equatorial rotation velocity of 2 kilometers per second.

Since the Sun is gaseous, however, its parts are not held together like those of the solid Earth. Instead, the rotation period lengthens by a degree that depends on the angle from the rotation equator to the pole, that is, on the solar latitude. At 60 degrees north and south of the equator, the rotation period is up to 28 days. The Sun thus rotates "differentially," the solar gases shearing past each other. Couple rotation with the oatmeal Sun, and something quite amazing happens.

Magnetism

Electricity and magnetism are an inseparable pair. An electric current, the result of electrons flowing in a wire, is always surrounded by a magnetic field; motion of a wire in a magnetic field in turn induces an electric current (the principle behind an electric generator). The solar interior is so hot that electrons are easily stripped from atoms, which become positively charged ions. These and free electrons result in electrified gases. Motion of the gases by convection and rotation then creates a natural magnetic field that the theoreticians tell us is made deep in the Sun at the bottom of the outer solar layer in which convection dominates the energy flow.

The differential rotation next stretches the field, and draws it into discrete strings and ropes. The outward pressure of the magnetic ropes makes the gases that hold them expand such that they become buoyant, and rise, taking the fields with them. When the magnetic ropes hit the surface, they pop through in loops rather like giant croquet wickets that are anchored to the solar surface. With strengths typical of medical MRI machines, the fields now retard the convection. Since less heat flows upward, the solar gases chill at the loops' twin bases. The result? Pairs of sunspots, one with a positive magnetic field, the other negative. Solar magnetism, however, is not created through stiff wires, but is free, and consequently unstable. So, then, are the sunspots, which typically live from a few days or weeks up—on rare occasions—to a couple months.

While the range of spot sizes is huge (from pinpoint to naked eye), their structures are all similar. At a spot's center is a dark, depressed basin that only appears black in contrast to the surrounding bright solar disk. Radiation from a hot gas or solid is terribly dependent on temperature. At 4000 or so Celsius, typical of spot centers, the gases release only 40 percent as much energy as they do at 5500°C. Seen alone, spot centers are actually quite bright. Sharply surrounding the center is a wide grayish ring that slopes upward to the normal solar surface and that is beautifully striated by the opening field. Multiple loops produce multiple spots, whole nests of spots, in which it is impossible to tell the pairs apart. As the nested fields dissipate or short-circuit, the spots disappear, only to be replaced by new ones until the whole collection dies away.

Windy Crown

At least for the next 100 million years the Moon will occasionally cross over the Sun to block the brilliant disk to reveal an extended outer atmosphere, a sort of solar "crown," the fabled corona. The Sun is so bright that only with highly sophisticated telescopes or from space (where the brighter blue sky does not get in the way) can we see the corona without the Moon's aid. Seeming to be about twice the size of the Sun itself, the pearly-white corona really has no discernible edge to it, but instead simply fades away, becoming ever fainter until it is lost in the background.

The solar corona radiates spectral *emission* lines (the reverse of the absorptions) at specific colors that have been identified with iron, nickel, and other atoms that have had half or more of their electrons stripped away, indicating violent collisions among them caused by fierce temperatures. These and powerful coronal X-ray emission tell of a very rarefied, but terribly hot, 2-million-degree gas (one so thin that it produces little radiation). Heat a gas and if it can, it will expand—a principle that ranges from steam engines to atomic bombs. Unless something holds it back. Even casual examination of the corona shows that much of it is sculptured into loops, suggesting that the coronal gas is at least partly trapped by intense magnetism pouring from the Sun and concentrated into the same looping structures that make the surface's sunspots.

While the magnetic loops retard the expansion, they cannot entirely halt it, especially in regions where they are weak or absent. The result is a hot "solar wind" that blows past—and slams into—the Earth at speeds of several hundred kilometers per second with temperatures well above 100,000 Celsius. The Sun, it seems, is losing mass, indeed is evaporating! No need to panic about losing our life-giving Sun, however, as the wind is so thin that at the current rate, it would take 100 trillion years for the Sun to "dry up." Far more dramatic things will happen to it when the internal hydrogen fuel runs out 5 billion years hence. So very tenuous is the wind that even were you in space you could not feel the high temperature nor have any sense that solar gases were tearing swiftly past you.

The severe laws of thermodynamics—the bane of all those who try (always unsuccessfully) to make perpetual motion machines—in essence say you can't get something for nothing, and indeed

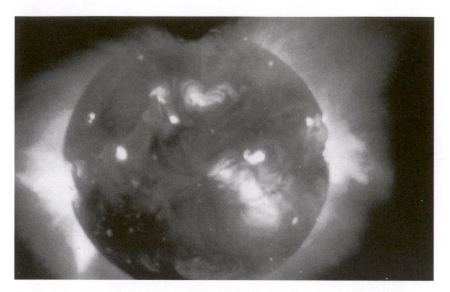

Figure 3.4. Observed from space by the orbiting Yohkoh Telescope, the hot corona radiates X-rays, allowing it to be seen against the face of the Sun, where the loop structures become even more dramatic. The solar wind blows very fast through the dark "hole" at lower left. Yohkoh X-ray Telescope, Lockheed-Martin, National Observatory of Japan, NASA/ISAS.

cannot even break even. A cool body cannot provide heat to a warmer one. So the corona cannot be heated to 2 million degrees by the radiation from the 5500-degree solar surface. The only available source is energy from the ubiquitous solar magnetism. Somehow, perhaps through waves in the magnetic fields, energy rides up the concentrated magnetic loops to deposit itself in the gas, the details far from understood.

What *is* understood is that the various loops can cross and reconnect with each other, and thereby violently collapse. Acting like giant atomic accelerators, the contractions slam atomic particles into the solar surface to create localized powerful "flares" that send out bursts of X-rays and that can be visible at all wavelengths, some so powerful that they are visible to the eye through the (properly filtered) telescope. At the same time, particles scream out in the other direction at nearly the speed of light to impact our local environment.

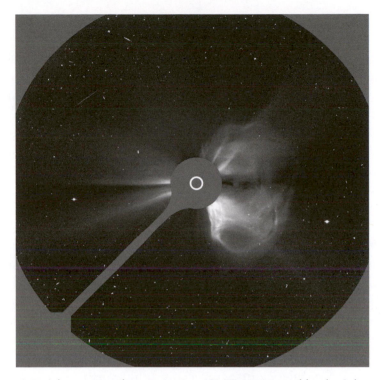

Figure 3.5. A huge coronal mass ejection (CME) is witnessed by the Solar and Heliospheric Observatory (SOHO), a spacecraft that orbits the Sun in front of Earth and takes daily images of the Sun and its corona. The long arm supports a disk that hides the bright solar surface, which is represented by the circle at center. At the peak of the solar cycle, the Sun can release several CMEs per day, though ones as big as this are much more rare. Distant stars flock the background. SOHO LASCO consortium, ESA and NASA.

The disappearance of a strong magnetic loop has an even more profound effect. Suddenly, the hot coronal gas that was confined by the field finds itself free of its magnetic leash, and does what a hot gas wants so badly to do: expand. The result is a "coronal mass ejection" (CME) that sends a bubble of corona screaming down the solar wind. CMEs, which are not clearly related to flares, can be spectacular and can involve nearly the whole Sun. They are also highly directional. If they miss the Earth, nothing happens, but if we are in the way, all sorts of things transpire, some quite beautiful, others quite expensive.

Return to Earth . . .

Like the outer envelope of the Sun, the Earth's hot liquid iron core circulates up and down by convection and is also stirred by rotation. Iron is an electrically conducting medium, and its movement again generates a magnetic field. Not concentrated into ropes as it is in the Sun, the Earth's field emerges from the interior as a sort of "dipole" rather as if a bar magnet had been shoved through the planet, one with a north magnetic pole near the north rotation pole, a south magnetic pole near the south rotation pole.

But there the simplicity ends. While the magnetic and rotation axes are more or less aligned, there remains a distinct tilt between the two of some 10 degrees. The north magnetic pole is in northern Canada, while the southern one is near the edge of the Antarctic ice sheet. Moreover, the locations tend to drift around by a few degrees. Since opposite magnetic poles attract and like ones repel (try it with refrigerator magnets), a floating magnetized needle—a compass needle—will align to the Earth's field and point the way and allow for terrestrial navigation. But sailor beware: it will take you toward magnetic north, not geographic north (or south), and a correction must be applied. (Such magnetic tilts are the rule. We see the phenomenon in the fields of the other planets, even in those of highly magnetized stars.)

Of much greater significance, the Earth's field is, unlike a simple textbook description, not at all ordered and symmetrical, or even constant. The electrified protons and electrons of the solar wind drag the Sun's field outward with them. Ramming into the Earth's Sun-facing field several Earth-radii out, the speeding wind sets up a gigantic shock wave (a magnetic version of a continuous sonic boom off the fuselage of a supersonic aircraft) and compresses it like a pancake. On the nighttime side, the wind blows the field into a huge extended tail—the "magnetotail"—that stretches past the Moon. Solar particles trapped by the Earth's field create two wide equatorial donut-shaped rings, the "Van Allen radiation belts," the inner one roughly 1.5 Earth-radii out, the outer one centered near four Earth radii, the whole structure becoming the Earth's "magnetosphere."

The interaction between the Sun and Earth sets up rings of electrical current that circulate around the magnetic poles only 100 or so kilometers above the Earth, with radii of about 20 degrees of latitude. The power of billions of watts strips electrons from atoms high (120 kilometers, 75 miles or more) in the Earth's atmosphere, causing the upper air to glow white, red, green, blue, depending on which atoms and molecules are being excited. The northern ring passes over Alaska, northern Canada, and northern Scandinavia. At night, from the ground, the sky glows with the spectacle of the aurora borealis, the northern lights. The aurora australis, the southern lights, similarly shimmers over deep southern climes where, sadly, there is hardly anyone (barring penguins) there to see them. Not confined to a wire, the current rings—like a huge electric spark—are unstable, resulting in spectacular auroral forms: hanging draperies, searchlight beams, sunburst coronae, all pulsing and rolling across the heavens.

. . . And Get Hit

We left the Sun with a CME pounding down the solar wind and aiming at Earth. Taking a day or two to get to us, the power of its strike grossly disturbs the terrestrial magnetic field, compressing the front side and stretching the magnetotail even more. Magnetic compass needles go awry. The current ring, bombarded by more particles, strengthens and expands toward the equator, allowing the more populated areas of Earth to admire the awesome beauty and power of the aurora, the Sun reaching out to touch our souls.

The Sun can also touch our ability to do science, our pocketbooks, even (under special circumstances) our health. Astronomers use telescopes for two reasons: to resolve fine detail and to observe faint sources of light that include distant Solar System objects, stars, and galaxies. But from the ground, where we find the biggest telescopes (the Hubble Space Telescope is not very large), our atmosphere impedes us. The most obvious effect is the twinkling of stars caused by variable refraction (bending) of light in turbulent atmospheric cells that makes celestial objects seem blurry. A secondary

Fig 3.6. Interactions between the solar wind and the Earth's magnetic field produce the aurora borealis, a common sight in the Alaskan and Canadian arctic and subarctic. Extra disturbances from coronal mass ejections allow aurorae to be seen in moderate latitudes far to the south (as here). The bright "star" toward the bottom is Saturn. The stars of Gemini lie above it. J. B. Kaler.

effect is a sort of permanent low-level aurora called "airglow" that places a limit on our ability to go as faint—and as distant—as we would like. Both are reasons for having orbiting space telescopes (in addition to the ability to access wavelength domains blocked by our atmosphere such as the more energetic ultraviolet and X-ray).

But in space, we have problems as well. Within the confines of the Van Allen belts, we are protected, as are all low Earth-orbiting satellites, which are only a few hundred kilometers up. Communications satellites, however, are far away. As the distance of a satellite from the Earth increases, the gravitational field drops, and its orbital period increases. Near the Earth's surface, satellites whip around with periods of an hour and a half. Place one at 35,800 kilometers (22,250 miles) from the Earth's center, however, and it takes 24 hours to make an orbit, the same as the Earth's rotation period. As a result, from the ground the satellite seems to hang motionless in the sky, that is, it becomes "geosynchronous."

At such distances satellites are beyond Earth's magnetic protection, and—along with numerous high-orbit research satellites—can be bombarded by solar particles, even electronically destroyed, wasting hundreds of millions of dollars. Early warning satellites that monitor "space weather" can tell of impending doom so as to allow controllers time to switch off sensitive electronics. Think then of the danger of human spaceflight. A CME smashing into a manned spacecraft on its way to the Moon or Mars generates X-rays that would present huge danger. The Apollo astronauts were lucky in that no major events erupted from the Sun. The dangers hold true for any human walking the sunward side of the Moon as well, since there is neither atmosphere nor a protective magnetic field.

Meanwhile, back on Earth, increased ionization of the upper atmosphere that occurs during an auroral event dramatically changes short-wave radio propagation (which uses electrified atmospheric layers as a radio mirror to bounce signals around the curvature of the Earth). At such times, we can pick up broadcast signals from anomalously far away. Radio amateurs might even talk to themselves, their signals circumnavigating the globe.

The electrical activity of the aurora can induce weak electrical currents to run in the ground, along the surface of Earth itself (though nothing you could sense). These in turn can cause chaotic currents to flow in long-distance power lines. Circuit breakers then pop, closing down systems. In the most famous case, a 1989 CME-induced aurora fouled up Quebec's power grid, resulting in a massive blackout and a huge cost. Similar events in the nineteenth century made a mess of telegraphy, one office even set on fire!

Even on Earth the protection to human health is not absolute. Solar particles riding down the field lines to the magnetic poles produce enough radiation to make high-altitude over-the-pole flights a concern, especially for the pilots and airline staff that do it constantly, and possibly even for pregnant women.

All planets with magnetic fields are subject to such disturbances. Aurorae have been seen on Jupiter and Saturn. The solar wind finally comes to a halt about 100 Earth-Sun distances out, where it crunches against the gases of interstellar space, to create a complex set of shock waves in a vast and mysterious region that the *Voyager I* spacecraft (launched in 1976) first began to crack in 2005.

Vanishing Act, Part I

Except when affected by solar events, the Earth's magnetism seems quite stable over the years, even centuries. Over longer periods, though, hardly anything seems stable, true of terrestrial magnetism as well. The deep rocky layer outside the iron core, while not liquid, is so hot that it behaves as a soft plastic, and like the outer core is subject to convection. Huge upwelling currents act to deposit new molten rock in a series of undersea ridges, the most obvious of which is the Mid-Atlantic Ridge that snakes between Europe/Africa and North/South America. As new crust is formed it pushes the old crust outward and away from the ridge, which in turn pushes the continents apart. Some 250 million years ago, Africa and South America were one land mass that were broken by a convective upflow. Not only do the shapes of these continents fit together, but so do various other geologic features. Europe and North America broke up as well. The pairs are still being driven away, measured at a rate of a few centimeters a year (oddly about the same rate at which the Moon recedes from Earth due to tides). Indeed, all the continental landmasses are shifting around like ships adrift on a mantle-sea of hot rock. Where they slam into each other, or override the oceanic crust (as on the west coast of the United States), we find great geologic activity that includes earthquakes and volcanoes.

When molten rock solidifies, it freezes into itself the direction of the local magnetic field. Residual magnetism in ocean rock nearest the Mid-Atlantic Ridge lies all in one direction parallel to the ridgeline. Move away from the ridge in either direction, however, and the magnetic directions all suddenly reverse. Farther away yet, they reverse again, forming clearly defined magnetic stripes. From the rate of seafloor spreading, these fossil magnetic remnants reveal that about every few hundred thousand to a million years, the Earth's field reverses direction! The cause seems to be chaotic circulation in the molten core.

In the reversal, the Earth's field can disappear for a time. If the field goes away, so do the protective radiation belts, which allows the solar wind—and the CMEs—to smash directly into the Earth's atmosphere. The resulting radiation damage to life forms may in fact promote genetic mutations and speed up evolutionary processes. The last reversal was several hundred thousand years ago. Over the past 150 years, the field strength has dropped by about 10 percent. Is this a normal fluctuation? Is it the beginning of a new reversal? Are we in trouble?

Solar Cycle

The solar magnetic "dynamo," true to its name, is dynamic and constantly changing, not just day-to-day, but over much longer intervals as well. While the individual spots come and go quickly, their numbers and positions tell of an average 11-year cycle, the "sunspot cycle." At minimum, we may see a lengthy "sunspot desert" with no spots at all. The beginning of a new cycle is marked by a few spots that begin to appear in both hemispheres about midway between the solar equator and the poles. As the years pass, more and more spots appear, but at lower and lower solar latitudes. Half a dozen or so years into the cycle the solar disk may be covered with dozens of them. Then as the average distribution of spots nears the equator, the numbers die away until practically none is left and we start anew at higher latitudes. All the magnetic phenomena that accompany sunspots partake in the 11-year cycle, including flares, coronal mass ejections, even aurorae, making the sunspot cycle really a magnetic activity cycle.

Though solar magnetism is concentrated in the spots, the Sun also exhibits a weak magnetic dipole reminiscent of that of Earth, with magnetic poles of opposite direction aligned along the rotation axis. In each hemisphere, the leader spot of a pair (that in the direction of rotation) will always be of one magnetic polarity (positive or negative), that of its relevant rotation pole, while the follower spot will take on the other polarity. In the opposite hemisphere, the polarities will be reversed. The magnetic directions stay the same during an entire 11-year cycle. During the *next* cycle, however, everything is switched, plus for minus, minus for plus. Only after 22 years are we back to the first magnetic configuration, the true magnetic cycle of the Sun double the length of the raw sunspot cycle.

The origin of this behavior is reasonably well understood. The solar differential rotation, which creates the magnetic ropes, wraps them ever more tightly over the 11-year period until they break down into tiny cells of positive and negative magnetism that are attracted to their opposite poles, where they establish a general field in the opposite direction. Once re-ordered, the wrapping begins anew with reversed polarities.

Effects

Solar activity bathes the Earth and its magnetic environment in high-speed particles, harsh ultraviolet light, X-rays, even gamma rays. Even though at cycle maximum the Sun is rich in dark spots, its total energy output actually goes up thanks to the increased short-wave emission, which (along with the particle emission) also has the periodicity of the sunspot cycle. The energy input to the Earth therefore involves much more than steady ordinary optical and infrared sunlight and the accompanying heat. All manner of things have been linked to the cycle and its variations, even the stock market. Many of the perceived cyclic links are the result of bad statistics, the Market a fine example, but some of them are very real.

The problem with uncovering the links involves these same shaky statistics. Take any two phenomena and graph them together and, if the numbers are small, there is a reasonable chance that they will accidentally correlate (one changing with the other), which can result in some really bad science. (Two points *always* correlate!) Only if one persists, and plots large amounts of data, can the correlation, if there is one, be uncovered. And even if true correlations are found, there is no guarantee that one phenomenon is the cause of the other and that they are not both subject to some unknown third variable element.

Except for scattered observations of naked-eye spots (seen in the dimmed rising or setting Sun), sunspot science began with the telescopic observations of Galileo in 1610. With only 400 years of data and a mere three dozen or so cycles, false relationships with terrestrial events are easy to make. There is, for example, no clear consensus on climate patterns that may respond to any particular cycle; any such effects are totally lost in chaotic variations in ocean currents and in the Earth's cloudy, churning atmosphere.

Yet the effects are present. Over a single cycle, the Earth's outer atmosphere heats and expands during sunspot maximum and contracts during minimum. There is enough residual air at the altitudes of low-orbiting satellites that atmospheric friction degrades their orbits, ultimately leading to complete de-orbit and collision with Earth. The Hubble Space Telescope is periodically boosted back to its original orbit through *Shuttle* missions. Satellites come down much faster at spot maximum than they do at minimum, a prime case in point being the 1970s' *Skylab* satellite, which crashed before the newly born *Shuttle* could put it back to where it belonged. The economic impact of the destruction of expensive satellites may in turn lead to a *real* stock market effect.

The cycle also enters in by way of the aurorae and related terrestrial magnetic disturbances. Aurorae are more common near and just after solar maximum, which leads to greater possibilities of power failures. The permanent airglow is also greater at sunspot maximum, which inhibits astronomical observations from the ground, the best observing slated for sunspot (magnetic) minimum.

More important may be the solar effect on high-speed atomic particles (mostly protons) from space called "cosmic rays" that we will encounter (safely) in chapter 7. They've been implicated in triggering cloud formation, even lightning strokes. Launched to us by exploding stars, the invaders are to some degree kept at bay by the great fort of the solar wind. Since the solar wind is modulated by the solar cycle, so then may be our cloud cover, and (though there is no direct evidence) lightning activity. The solar cycle may in this way help modulate the weather, though the relations are so complex that we are far from understanding them.

Longer periods seem to have an effect as well. Solar cycles do not repeat themselves very well from one to the next. The 1958–59 peak was magnificent, as were the accompanying aurorae, while the next one around 1970 was a bust. The following two were sort of in-between. The level of cycles to come seems to bear a relation to solar action over the past few cycles, and is (to the relief of those operating expensive satellites) possibly becoming somewhat predictable. The interval between cycle maxima apparently also varies over an 80-or-so-year period. When the cycle shortens down to 10 years and under, the Earth heats slightly, while a longer set of cycles seems to cause the Earth to chill a bit, which may well lead to still-unrecognized climate changes. No one knows why.

At least one thing that you can rest easy about is the effect of solar activity and the solar cycle on sunburn. Sunburn (which, with its attendant skin cancers, you *do* have to worry about) is produced by the "near ultraviolet," the part of the ultraviolet spectrum that falls just to the short-wave side of ordinary violet light. Harsher ultraviolet, that linked to magnetic activity, is screened out by the Earth's atmosphere. Indeed, there is some evidence that the Earth's air becomes a better ultraviolet barrier during maximum. It's still not a good idea to sunbathe, as Sun can do more than reach out, it can truly get you. And for sure, don't do it in space!

Vanishing Act, Part II

Over still longer periods of time, however, the solar activity effect becomes dramatic. Shortly after Galileo and others began solar observations, sunspots effectively disappeared. Between around 1645 and 1715, hardly any were noted at all. This long-term "Maunder Minimum" (after the English astronomer Edward Maunder) coincides with a distinctive period of intensified cold weather called the "Little Ice Age," when northern Europe was covered with deep snows, the Norse colonizers were frozen out of Greenland, rivers that had never frozen iced up, and so on.

Again, two events happening together hardly make a convincing cause-and-effect. However, even though the spots themselves cannot be tracked prior to 1610, other effects of solar activity *can* be traced to yield information on the solar cycle. A radioactive isotope of carbon, C-14 (6 protons, 8 neutrons, giving 14 nuclear particles), is produced by collisions between nitrogen atoms high in the air and (here they are again) cosmic rays. Living things incorporate the C-14, which then proceeds to decay away. Since cosmic ray influx depends on solar activity, measurement of C-14 in ancient tree rings and in layers in deep ice cores then provides a measure of solar behavior that can be related to the historical climate record. Old reports about auroral activity add to the database.

The "Spörer Minimum" of 1550 corresponds to another period of chilling, while between AD 1100 and 1400 the northern hemisphere appears to have warmed in accordance with an activity level similar to and even higher than we see today. Which is why the Norse were able to colonize Greenland in the first place. (If there *had* been a stock market . . .) Earlier minima are identified as well. There seems to be little doubt that the solar cycle has a powerful long-term effect, which confuses issues (including the politics) of global warming, which in turn affects the economy, and we are back to the stock market.

Vibes

The connections may be more intimate than we had ever realized. The solar cycle is a gentle long-term affair, while coronal mass ejections and flares are quick and violent. In between are solar vibrations. Convective motions and solar flares, as well as other motions, set the Sun vibrating, making it ring like a bell. The principal oscillation has a period of around 5 minutes, but there are myriad shorter-period overtones as well as longer oscillations. The oscillations allow us to measure the solar structure, chemical composition, internal rotation, even age (which agrees with that from meteorites). Moving now from concrete knowledge into speculative territory, there is some evidence that these vibrations can ride the solar wind, where they somehow couple with Earth to cause low-level seismic waves within the Earth's crust. As the Sun rings, so does Earth. Related magnetic effects have even been implicated in communications disturbances.

So next time at the beach, contemplate not just the eternal tides, but (while wearing sunscreen to protect against ordinary solar ultraviolet light) also contemplate that the Sun reaches out to touch us in unexpected ways through its wind, convection, silent ponderous rotation, magnetic field, even vibrations. Everything, it seems, is connected to everything else. Even the distant planets can affect one another, and thus us, leading perhaps to bigger chills than a diminished solar cycle could ever begin to produce.

Chapter 4

Frozen Earth

Ice. In typical temperate winters, it comes and goes with the seasons. Only in the far north and south (and on tall mountains) is it permanently installed, huge amounts of it surrounding the cold rotation poles and blanketing Greenland and Antarctica with thick glaciers. As great as they seem, these ice caps are but ghosts of what you would have seen 20,000 years ago during the last ice age, when the northern one stretched well into the central United States, a time when woolly mammoths and other now-extinct beasts walked life's stage, and primitive men and women used stone tools.

This period of deep natural refrigeration was not unique. Like the current annual wintery cycle, ice ages come and go. Here are not "little ice ages" caused every few hundred years by the relaxation of solar activity, but true ones that over intervals of 1000 centuries plunge parts of the world into bitter cold, their vast ice sheets changing and scouring the land over which they ebb and flow. You can see their effects in the great valley of Yosemite, in the basins of the Great Lakes, in the fertile soils of the American prairie. The origins of ice ages have been debated, argued, fought over, with no consensus. One explanation is astronomical, that they were brought to us by periodic changes in the Earth's orbit and orientation, which in turn are the products of the gravitational interactions between our planet, the Sun, and all the other bodies of the Solar System as they reach out to touch us and make us pay them heed. To understand the long-term trends, however, we need first look more closely at the short.

Figure 4.1. The icy glaciers of Antarctica give a sense of what northern temperate latitudes must have looked like during the many ice ages that covered the land as far south as central California and Illinois. Courtesy of Caroline and George Badger.

Seasons

Since deep winter in the Earth's northern hemisphere coincides with the closest approach by the Earth to the Sun (perihelion, about January 2), and since winter and summer alternate between terrestrial hemispheres (when it is winter in the north, it is summer in the south), variation in solar distance can have little to do with the seasons. Their origin instead lies in the Earth's orientation.

Our planet rotates in a day, revolves about the Sun in a year. As Earth spins from west to east, the distant stars (millions of times farther than the Sun) appear to move oppositely, rising in the east, traversing daily paths, and setting in the west. As we orbit, the Sun appears to move to the east against the stellar backdrop, in the direction opposite to that of its daily path, but much more slowly, at but a degree per day, year after year tracing out its ecliptic path, which defines the famed star-patterns—constellations—of the Zodiac: Aries (Ram), Taurus (Bull), Gemini (Twins), Cancer (Crab), Leo (Lion), Virgo (Maiden), Libra (Scales), Scorpius (Scorpion), Sagittarius (Archer), Capricornus (Goat), Aquarius (Water Bearer), and Pisces (Fishes). That all but one are living things is testimony to the sacredness of the Sun and of ancient astrology (and Libra, as part of Scorpius, was once "alive" too).

Rotation is uncoupled from revolution. Rather than an even number of days in the ordinary year, there are 365.24219 . . . , the number's decimals unending (which makes calendar construction interesting). Just as important, the rotational axis of the Earth is not aligned with that of the orbital axis, the perpendicular to the Earth's orbital plane (the plane of the ecliptic), but tilted to it by an angle of 23.4 degrees. This number—the "obliquity of the ecliptic"—is but an accident of nature, of the formation and history of the Earth. All planets have such tilts that range from near zero for Mercury, through a strange 98 degrees for Uranus, to 177 degrees for Venus (the latter two thus rotating backwards).

As the Earth has rotation poles, so does the sky, all celestial objects seeming to move along daily paths that are centered on the North and South Celestial Poles (which lie directly above the Earth's north and south poles). In between, the sky's equator—the celestial equator—lies above and parallel to the Earth's equator. In the northern hemisphere, the north celestial pole (marked nicely by the star Polaris) is above the northern horizon, while in the southern hemisphere, the south celestial pole is up. The celestial equator, splitting the difference between the poles, runs across the sky from due east to exact west.

Since the rotation and orbital axes are inclined to each other by 23.4 degrees, so are the ecliptic and celestial equator. Each of the two circles is centered on the Earth, and therefore must intersect at two points, again at an angle of 23.4 degrees. As the Sun moves along the ecliptic, it must cross the equator twice a year, once on its way going north on March 20, once on its way south on September 23, dates that should ring bells, as they define the first day of spring (in the northern hemisphere) and the first day of fall. Since the sky's equator goes from east to west, on those dates the Sun rises in the exact east, sets exact west, and is overhead at the Earth's equator. Since just half the equator is visible, days and nights are equal at 12 hours apiece, hence the intersections' names: "equinoxes," vernal at the March 20 crossing, autumnal at the September 23 intersection.

The 23.4-degree tilt of the ecliptic allows the Sun to plunge as far south as 23.4 degrees below the equator on December 22, the first day of northern winter, at the "winter solstice," and to climb as high as 23.4 degrees above the equator on June 21, the first day of northern summer, at the "summer solstice." On June 21, the Sun shines overhead at 23.4 degrees north latitude at the Tropic of Cancer, and overhead at 23.4 degrees south latitude at the Tropic of Capricorn, the symmetry beautiful to behold.

Here is the origin of the seasons. When the Sun is high in the sky, sunlight falls more nearly perpendicular to the ground and has maximum heating ability: which is why northern-hemisphere vineyards are often on the south-facing sides of hills. When it is low in the sky, as it is all day in northern hemisphere winter, sunlight hits at a sharp angle to the ground. It is thus forced to spread itself out over a larger area, and the local Earth chills. The hottest places on Earth are where the Sun is most nearly overhead, between the two tropics, and coldest where it never climbs very high, near the poles, in spite of the fact that above the Arctic Circle at 66.6 degrees (90 − 23.4) north latitude, and below the Antarctic (66.4 degrees south), it can be up for 24 hours straight.

The variation in distance caused by the elliptical orbit adds a small modulation onto the seasonal cycle. Very much like the strength of gravity, the amount of energy we receive from the Sun depends on the inverse of the square of the distance. Since the variation in distance is 3.4 percent, the Earth receives 7 percent more heat energy at perihelion than it does at aphelion (where Earth is farthest from the Sun). Our relative closeness in northern-hemisphere winter then adds a little heat to the frozen land, while the increased separation in northern summer is a bit of an aid to air-conditioning. In the southern hemisphere, the distance effect is reversed, in principle making the seasons somewhat more extreme than those in the northern hemisphere. Much of the effect is lost,

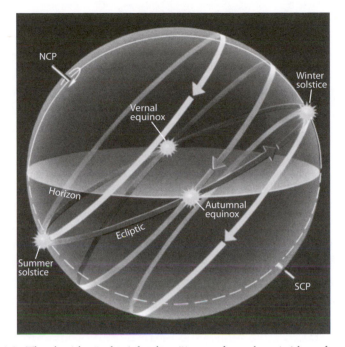

Figure 4.2. The sky (the "celestial sphere"), seen from the outside as from a mid-northern latitude (with Earth and you at the center), is captured in time. From our seemingly stationary perspective, it rotates clockwise around the North and South Celestial Poles (NCP and SCP), with the unlabeled celestial equator in-between. The daily path of the Sun—and the seasons—depends on the Sun's location as it rides annually along the tilted ecliptic. When at the equinoxes, the Sun rises and sets due east and west. If at the winter solstice, it rises in the southeast, sets in the southwest, and crosses low above the horizon, resulting in minimum heat absorption by the ground; if at the summer solstice, it rises and sets in the northeast and northwest, and at mid-day is high, resulting in maximum heating. (The equinoxes are shown rising and setting. As the Earth rotates, they too move across the sky, resulting in varying orientations of the ecliptic relative to the equator and horizon.) From *Stars*, Scientific American Library, New York: Freeman, 1992, © J. B. Kaler.

however, in the distribution of land masses, most of which are in the north, leaving the south with its great oceans. Water, which is a huge heat reservoir, moderates the southern seasons, making the distance effect unnoticeable.

Yet there is another "however." Though our planet is controlled largely by the mutual gravity between it and the Sun, there are other factors in the Solar System. Each planet (and the Moon, as we saw), has its own gravitational field, and each pulls on us. Over long time periods, the Earth changes not just its orientation in space, but its actual orbit, which in principle could lead to profound changes in global and local climate.

Perhaps even to ice ages.

Wobbles

The most potent effect of lunar and solar gravity—after the orbits themselves—is the perpetual ocean tide. Following that is another that comes into play only over lifetimes are strung together. Spin a child's top, or a gyroscope. While it quickly rotates, it also is subject to wobbling, its axis spinning madly around. Only if there are no outside influences will the wobble—the "precession"—be absent. And the Earth certainly has plenty of these.

Terrestrial rotation bulges the Earth out at its equator by about 40 kilometers (25 miles) relative to the axial diameter. The Sun, on the other hand, is always found along the tilted circle of the ecliptic, while the Moon must be close by, always within 5 degrees of the ecliptic. Only when the Moon and Sun are at the equinoxes will their gravitational pulls lie along the bulge's plane. At all other times, lunar and solar gravity pull either up or down on the bulge in an attempt to lower the tilt angle of the rotation axis. Rotation, however, counters the effort by imparting great stability, the reason a navigational gyroscope points in a constant direction.

The result is that, instead of being pulled "upright," the Earth's axis undergoes a slow rotation, a wobble, in which the rotation poles each inscribe small circles of 23.4 degrees radius around the perpendicular to the orbit. With a period of 25,800 years, precessional motion would seem inconsequential. Yet the effect is big enough that it was discovered with naked-eye observations, by the great Greek astronomer Hipparchus of Nicea, around 150 BC!

While the forces generated by precession are not sensible to any instrument, we can see what is going on by looking at the sky. There are numerous related clues. The most notable is the change

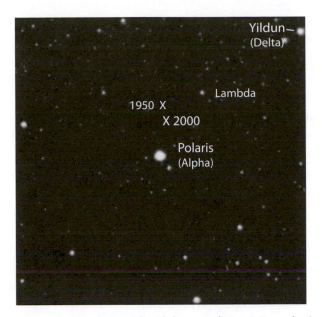

Figure 4.3. Polaris, the North Star (the Alpha star of Ursa Minor, the Smaller Bear), lies at the end of the handle of the Little Dipper. The next star along the handle, Yildun (Delta), 4 degrees away, is at the upper right corner. The big "X," 3/4 of a degree from Polaris, locates the sky's north rotation pole, the smaller "x" its location in 1950. Though for a while Polaris's position will improve, we will eventually have to start looking toward a new pole star. J. B. Kaler.

in the north or south "pole star," the star closest to one of the Celestial Poles. Right now, Polaris, the North Star, occupies that lordly position. That the Earth's northern axis points at Polaris (or nearly so; it is three-quarters of a degree off the true pole) is a coincidence. Although it is the Earth that is really precessing, since we are on the moving body, it looks instead as if the sky is rotating past the pole. As the axis precesses, Polaris will appear to draw closer to the pole until about 2105, when it will be only a quarter of a degree away. The two will then separate. In 13,000 years, the pole will point 47 degrees (twice the obliquity) *away* from Polaris. In 2700 BC, during the time of ancient Egyptian civilizations, Thuban in Draco was the pole star, while in AD 12,000, brilliant Vega will be at least fairly close. And in 26,000 years we get Polaris once again.

If the pole wobbles, so does the equator—of both the Earth and sky. If the equator wobbles, the intersections with the ecliptic must change their positions against the stellar background (hence the proper name for the wobble, "precession of the equinoxes"). Right now, the vernal equinox lies in the direction of the constellation Pisces, but it was not always so. Some 3000 years ago, the equinox was in the next zodiacal constellation to the east, Aries, the reason for that figure topping newspaper horoscope lists. In a couple thousand years, the equinox will have moved into Aquarius, and then will travel all the way around until, in 26,000 years, it is back in Pisces. Most astrologers do not allow for precession, and keep the sign of Aries stuck to the vernal equinox. As a result, the signs of the Zodiac more or less overlap the next actual constellation to the west of where they once were.

Similarly, the autumnal equinox (now in Virgo) was once in Libra (the "balance" of days), the summer solstice (always between the equinoxes and now on the Gemini-Taurus border) was once in Cancer (giving us the tropic of Cancer), and the winter solstice (now in Sagittarius) was once in Capricornus (hence the tropic of Capricorn). However, like the celestial poles, the equinoxes and solstices appear fixed to us on Earth. Since we do not feel our motion, it *looks* as if the constellations are sliding past these points as if on a moving belt. In half the precessional cycle, northerners will see Sagittarius high in the sky where Gemini is now, and Gemini will glide just above the southern horizon. It takes nearly 26,000 years for the vernal equinox to move through 360 degrees. Divide one number by the other, and find that the equinox moves 50.4 seconds of arc per year (a second of arc is 1/3600 of a degree), and thus a full degree in a human lifetime, more than enough to be recognized.

Precession gives rise to an additional definition of the year. How long it takes the Earth to go about the Sun depends on the point of reference. Our calendar year (the "tropical year," from the Greek for "turning") is the interval between successive equinox passages by the Sun, and is 365.24219 . . . days long. But the equinox is moving backwards (westward) against the stars. The "sidereal year," the interval between successive passages of the Sun relative to a given star, has to be longer, 365.25636 . . . days, a difference of 20 minutes! The sidereal year is the "true" year, the one that would be measured from the outside by chapter 2's Martian, and is the one that must be used in orbital calculations.

As if all that is not sufficient, since the Moon and Earth do not exactly share an orbital plane, the combined forces of the Sun and Moon on the bulge do not sum up to a constant amount, but continuously vary. In response, the equinoxes do not move smoothly against the background, but jitter back and forth by a small fraction of a degree with multiple periods, the most prominent being one of 18.6 years, which is the period in which the Moon's *orbit* precesses thanks to the effect of the Sun. The secondary effect, called "nutation," first noted over 200 years ago, helps make timekeeping a real trial.

The Moon, though a relative lightweight, is nearby, and the Sun, though far away, is massive, accounting for their strong influences on us. The other planets are neither close nor massive, and we might think to neglect them. But though their pulls are small, time is on their side, and the resulting effects important to our story. First, though, we need to introduce them.

Planets

Eight, nine, ten, eleven, maybe more, depending on how you define and count (the Pluto controversy apparently to be with us for a long time). The Earth, the third one out, provides the reference, its distance from the Sun 150 million kilometers, or one "astronomical unit" (AU). Traditionally, the planets are (with average distances, diameters, and masses):

Mercury (0.38 AU out, 0.38 Earth diameter, 5.5 percent Earth mass)
Venus (0.72, 0.95, 0.81)
Earth (1.0, 1.0, 1.0)
Mars (1.52, 0.53, 0.11)
Jupiter (5.2, 11.2, 318)
Saturn (9.5, 9.5, 95)
Uranus (19, 4.0, 14.5)

Neptune (30, 3.9, 17.1) and (let's keep it)
Pluto (40, 0.18, 0.002)

They can be grouped in different ways. Mercury through Saturn
are the "ancient planets," those known since prehistoric times. The
last three were discovered telescopically, in 1781, 1845, and 1930.
The inner two (those "below the Sun") are the "inferior planets,"
the outer six the "superior." As important, since Mercury, Venus,
and Mars are iron-filled rock balls like Earth, the inner four are the
"terrestrial planets." The next quartet forms two pairs of similar
bodies. Large cloudy Jupiter and Saturn are made mostly of hydrogen and helium (quite like the Sun), while the smaller near-twins
Uranus and Neptune, although loaded with these light elements,
contain a lot of heavier water and similar substances. The big outer
four planets have numerous satellites, while the inner four do not
(and include just our Moon and two tiny Martian ones).

The first eight planets orbit the Sun in more or less circular orbits in the same direction and pretty much in the same plane. Such
symmetry led to the idea that the planets were accumulated from
a disk of dusty gas that encircled the primitive, just-born Sun 4.6
billion years ago (the age derived from the amounts of decaying
radioactive elements found in rocks). Orbiting mostly between
Mars and Jupiter are the smashed remains of what might have
been a planet had Jupiter's gravity not kept them from gathering
together—the asteroids—while outside the orbit of Neptune is an
extended disk of junk that is too thin to have been allowed to
form one in the first place—the *Kuiper Belt* that extends at least to
55 AU and probably farther. And then there is Pluto, a small ball
of icy rock with an odd, elongated, tilted orbit that has for decades
been taken as the "last planet," but is actually one of the Kuiper
Belt's largest and nearest bodies.

The smaller Kuiper Belt objects are dusty iceballs that, if they
wander too close to the Sun (the result of the gravitational action
of Neptune and the other planets), begin to evaporate, which frees
dust and small bits of rock. The released gas and dust are wafted
away by the solar wind and sunlight into streaming tails to form
comets. Surrounding the whole affair and extending toward the

nearest star is the *Oort Cloud*, a spherical volume filled with another trillion or so dark cold comets. Asteroids and comets have their own chapters. This one belongs to the planets.

In Motion

Planets move, the very name from Greek meaning "wanderers," in response to solar gravity. They orbit the Sun counterclockwise (as seen looking down from the north) with periods that depend on their distances, from 88 days for Mercury to 249 years for Pluto. As we watch from our shifting platform on Earth, we see them moving—some quickly, others slowly, some forward to the east, others seemingly backwards—against the background of the constellations of the Zodiac, not far from the apparent paths of the Moon and Sun.

For most of history, everyone thought they all went around the Earth. The breakthrough came in 1572 when Copernicus demonstrated that the apparently irregular motions made a lot more sense if they, including Earth, all went around the Sun. Based on careful observations by the Danish astronomer Tycho Brahe, between 1609 and 1620 Johannes Kepler worked a trio of rules that describe planetary motion: that planets move in elliptical orbits with the Sun at one focus; that they speed up when getting closer to the Sun and decelerate when getting farther away; and that relative to Earth, the squares of the orbital periods in years equal the average orbital radii in Astronomical Units cubed.

He had no idea why. That issue was resolved in the latter 1600s by Isaac Newton from a combination of his three laws of motion, his law of gravity, and his invention of calculus, with which he brilliantly re-created Kepler's laws, showing once and for all that gravity was the force behind the whole affair. But his re-creation, coming from more general principles, greatly expanded the scope of Kepler's simple laws. Newton showed first that orbits need not be closed, but could be one-way, that a body could approach Earth, swing around it, and go back out never to be seen again. He also revealed the second law to be no more than what we now call the

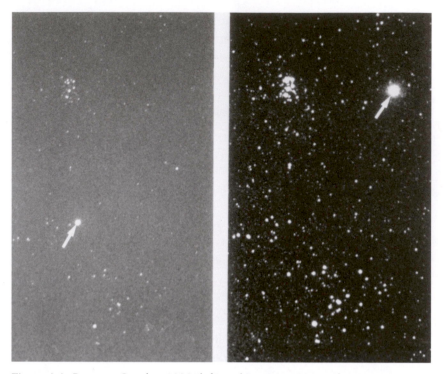

Figure 4.4. Between October 1988 (left) and January 1989 (right), Jupiter moved by a full 10 degrees to the west against the background of the stars of Taurus. The Pleiades, or Seven Sisters, are at the top, bright Aldebaran at the bottom. Planets generally move easterly against the stars. Here, Jupiter is moving westerly, or "retrograde," as the faster moving Earth swings in between the planet and the Sun. The images are identical, except that the photo on the left was taken from a city suburb, the one on the right from a dark mountain top. Look at what we have lost! J. B. Kaler.

conservation of angular momentum, which is so important in dealing with the effects of the tides.

The third became a more complex relation that involved not just the orbital periods and dimensions, but also the masses of the two bodies (which makes sense, since the force of gravity is mass-dependent). Equally significant, the new work was able to put these quantities in physical units: grams, centimeters, and seconds and the like. In the generalized third law, the period of a planet (in seconds) squared equals a known constant multiplied by the orbital size (in centimeters or meters) cubed, but now divided by the sum of the masses of the planet and Sun in grams or kilograms.

As before, the bigger the orbit, the longer the period. But now we see that the period also responds to mass. The greater the mass of either body, the greater the gravitational connection, and in response, the planet "falls" faster, which shortens the orbital period. Kepler got away with his simple law because the planetary masses are so small compared with that of the Sun that the sum of the mass of the Sun and any planet is—given the accuracy of naked-eye observations—essentially constant.

The power of Newton's variation on Kepler's third law is immense. It applies not just to the planets and Sun, but to *any* two bodies *any*where, be they planets and their moons, two stars in mutual orbit, even paired galaxies. Among the most important parameter for any astronomical body is its mass, which can be found if the body is in some kind of orbit with another. All one need measure is the orbital period and size, and the sum of the masses of the two bodies drops right out. If the mass of one body is obviously very small compared with that of the bigger one, then the sum is effectively that of the bigger body. A solution of the third law for the Earth gives the sum of the mass of the Sun plus the mass of the Earth, which since the terrestrial mass is trivial, is just the mass of the Sun. The same holds true for satellites in orbit about planets. If the orbiting masses are more or less comparable, we observe how each influences the other, that is, we can find the center of mass (about which they both move), which leads to the ratio of masses and thus directly to the individual masses. The "orbit" need not even be closed for the technique to work. Observations of the deviation of a passing planetary probe will do the job just fine.

Newton's legacy allowed no less than today's Earth-orbiting satellites and the exploration of the Solar System.

The Celestial Mechanic

Who, one might think, spends his time fixing the mechanism of the celestial sphere, oiling it when necessary, or replacing gears.

The tools, however, are not hammers and wrenches, but pencil, paper, and computers, as it is the celestial mechanic's job to *use* the known mechanisms of the laws of motion and gravity to predict the movements of the planets, indeed of all orbiting bodies, whatever forms they may take.

Stick for now with the Solar System, and with the classic example of two bodies (all other influences ignored). The mechanic begins with the observations of the movement of a planet. From observed positions and the basic laws, he or she determines the nature of the orbit, which involves the calculation of seven numbers that fix it in space relative to the Sun and to the orbit of the Earth. The mechanic must take into account only that the focus of the orbit is really the center of mass of the planetary system, not the Sun itself. Jupiter alone, for example, pulls the Sun off the focus by a full solar radius.

The first number is orbital size, the orbit's "semimajor axis," which is half the length of the longest line that can fit into the orbital ellipse, and which is also the average distance between the planet and the Sun. Second is the eccentricity, which mathematically sets the flattening of the ellipse, and which in turn tells the actual distance of the planet and its orbital speed at any position along the ellipse. Third is the *inclination*. Planetary orbits are not all in the same plane, but have their own peculiar tilts relative to the ecliptic. The Moon's path, for example is inclined by 5 degrees, Jupiter's by 1.3 degrees, Mercury's by 7 degrees, and Pluto's by a remarkably high 17 degrees (which can easily take it right out of the Zodiac).

The next two numbers involve the orbit's orientation. Since the planet's path is tilted, it must intersect the ecliptic at a pair of opposite points called the "nodes." The first number is an angular measure of where the nodes are relative to the equinoxes. And since the orbit is elliptical, it must have a perihelion, where the planet is closest to the Sun (and an aphelion where it is farthest). The second orientation number gives the angle between the location of perihelion and that of the node. To complete the set, we need only know the orbital period and a moment in time that tells when the planet actually passed perihelion. We can then calculate the planet's location for any time into the future or back into the past, and know where it was or will be against the stellar background.

Except that it doesn't quite work that way.

Planets Perturbed

The Earth's orbit (actually, the orbit of the Earth-Moon system) is dominated by the Sun. But the other planets have gravitational fields too that must somehow affect us. Their degrees of influence at any moment depend on their masses and on their distances from Earth. Venus, for instance, with a mass 1/400,000 that of the Sun, at its closest to us (0.28 AU) exerts 1/32,000 the solar gravity, giant Jupiter 1/18,000. While these numbers are small, the effects are steady and cumulative. Each planet pulls on Earth—and every other planet—with a different force and in a different direction, causing any particular planet to deviate slowly from the simple elliptical path based on it and the Sun alone. The planets even influence each other through their pulls on the Sun, as each shifts the solar location relative to the center of mass of the system. The calculation of an orbit for just two bodies is simple, and yields readily to mathematical solution. Add a third body and there *is* no explicit solution. You the mechanic must instead make do with approximate stepwise calculations.

Ideally you know the masses of all the planets from their satellites (either natural or man-made), or from observing the one-way orbits of passing spacecraft. At any given time, the locations of all the planets are known through observation, which gives their immediate distances and directions from the Sun and from one another. To find their real orbital paths, you first calculate all the mutual forces, hence accelerations, including those between the planets and the Sun. Then you calculate the direction, speed, and distance through which each planet (and the Sun) is forced to move over an adopted time interval, say a day, which gives a new set of locations. Do it all over again for the next day, and then the next, and so on. Of course the forces are not constant over a day, so you accumulate a bit of error that each day gets worse. To see how well you are doing, go to a half-day interval or less and see if you get a similar answer.

Given nine planets and the Sun, there are 45 forces to be calculated for the repositionings of ten bodies. If you want to calculate orbital positions over 10 years and pick an interval of a day, you have to run the calculation 3652 times. The job is clearly ready-made for high-speed digital computers. In days BC ("Before Computers"), the workload of orbital calculation was so huge that astronomers had to make do with approximate "perturbation formulae" that could at least give decent ideas of the deviations from an original "two-body" path. Their achievements were extraordinary.

God of the Sea

In 1781, just five years after the North American colonies broke with England, William Herschel, a musician and telescope maker from Germany who made England his home, broke open the traditional Solar System by discovering a "new planet," the first one known beyond Saturn. At 19 Astronomical Units from the Sun, double Saturn's distance, Uranus is just barely visible to the naked eye. Indeed, it had been seen a number of times before Herschel's great discovery, but had been recorded simply as a faint star. Herschel, however, had telescopic advantage, and could see that the "star" presented a tiny disk. The object's motion over the next year gave it away. Equally important, the motion led to the discovery of Neptune (and then indirectly to that of Pluto), in one of the greatest astronomical achievements of all time.

Uranus's orbit was quickly determined from observation, and its pre-discovery positions calculated using the planets known at the time. Oddly, Uranus was not quite where it was supposed to be. Continuing post-discovery observations showed the same thing, Uranus—the god of the sky—misbehaving in his own bailiwick. The only rational explanation that emerged was that the planet had to be influenced by another that had yet to be discovered. If there was a planet beyond Saturn, there may well be another beyond Uranus.

To find it, John Couch Adams in England and Urbain Leverrier in France used a new discovery technique. Since the basic laws were all known, they set themselves to calculate where the missing planet would have to be to produce the observed orbital deviations. The job was mathematically brutal. A common practice in elementary algebra is the solution of two equations with two unknowns, which takes a few minutes to run off by hand. Three equations with three unknowns may take half an hour, four times four, a full day. Here is 9×9, a task that took years. By the mid-1840s, each of the astronomers had a solution.

The English, however, failed to find the planet at its predicted position (the subject of a whole book in itself). Leverrier was done with his extraordinary calculation in August 1845, and in September transmitted the predicted position of the suspected trans-Neptunian planet to the Berlin Observatory, where they had new and accurate charts such that the planet might be found against the background of the stars. On the first night of observation, September 23, 1845, Johannes Galle put his eye to the telescope, and there it was, Neptune, the god of the sea rising from invisibility.

The discovery caused a sensation. Neptune was not discovered so much in the sky by Galle as on paper by Leverrier (Adams usually sharing the discovery) through calculation. Newtonian mechanics worked brilliantly. The Universe was eminently predictable! If one knew the positions and velocities of all particles at any particular time, then they were known forever. Even humans might be subject to Newtonian law. There was no free will! ("Your Honor, I was compelled to rob the bank by forces beyond my control. . . .")

Even with Neptune in the equation, Uranus still refused to behave itself. Was there yet another? A similar search was launched that culminated in the discovery of Pluto in 1930 by Clyde Tombaugh. It was, however, a tiny thing in a wacky orbit that could not possibly have enough mass to affect Uranus. Instead, Pluto's

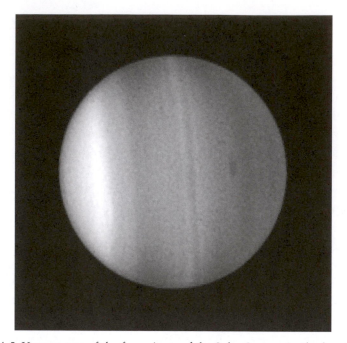

Figure 4.5. Uranus, one of the four giants of the Solar System, is tilted on its side. Over the past two decades, as the Sun has moved from near overhead at the south pole (to the left) toward the equator, the planet has developed a rich weather system, including a dark "storm" to the right. The Sun crossed the Uranian equator at the "vernal equinox" on December 7, 2007. The seemingly erratic motion of Uranus was instrumental in finding Neptune. Hubble Space Telescope: NASA, ESA, L. Sromovsky, and P. Fry (University of Wisconsin), H. Hammel (Space Science Institute), and K. Rages (SETI Institute).

discovery had been "accidental," except that years of work by Tombaugh had hardly made it an accident. For additional years a search was continued for "planet X." The problem was not "X," however, but correct planetary masses, which, when finally known through spacecraft observations, told us that the Solar System—even Uranus—was in proper order.

Nevertheless, Pluto opened our eyes to the wider system. It is not so much a planet as the first known body of the Kuiper Belt, which feeds us with short-period comets. Some 1000 objects are now known in the frozen realm of the Kuiper Belt, at least one ("Eris") bigger than Pluto as well as others not much smaller.

Changing Earth

No matter where we might live, we take comfort in the reliable changing of the seasons, of the familiarity of the weather patterns over the years. But move the Earth via the gravity of other planets coupled with the precession, and the amount of sunlight falling at any particular location must change with time, as—presumably—must the climate. The changes are so slow that over the short term of a human lifetime the effects are hardly noticeable. Are the ice ages the record of the long-term effects?

Precession by itself has no effects on the seasons or climate. It merely changes the view, the stars that we see overhead or during any given month. But now couple it with the eccentricity of the Earth's orbit. Orbital eccentricity causes southern hemisphere seasonal variations to be somewhat greater than those of the northern hemisphere, an effect that is "watered down" by the southern hemisphere oceans. In 13,000 years, though, the axis will be tipped opposite to its current direction, such that perihelion will take place in northern summer rather than winter and aphelion in winter rather than summer. The modulation of the seasons caused by the eccentricity will therefore be reversed, making northern hemisphere seasonal variations greater, and southern hemisphere variations more moderate. Since the south is moderated anyway by oceans and the north not so much, the effect will be notably exaggerated.

However, and there always seem to be "howevers," we for the moment forgot about the planets. Their pull on the Earth slowly causes the orbit—perihelion and aphelion—to rotate forward within its plane, in the same direction in which the Earth orbits. So far we have two "years," the tropical and the sidereal, the difference caused by precession. Now we have a third, the "anomalistic year," the interval between two successive solar passages of perihelion, which is the time it takes the Sun to make a full circuit of the ellipse. Since perihelion is advancing forward, it takes the Earth longer to catch up with it than it does with the equinox or a star, making the anomalistic year longer than either of the other two, 356.25964 . . . days, 25 minutes greater than the tropical year and 5 minutes greater than the sidereal.

Relative to the stars, perihelion advances to the east by about a third of a degree per century. All things being equal, the elliptical orbit therefore makes a full 360-degree rotation in 1080 centuries, or 108,000 years. But our calendar year is based on the precessing equinox, which is going much more rapidly in the other direction, clockwise. Since perihelion is going counterclockwise, the equinox catches up to perihelion before it catches up to any particular star, giving a shorter precessional period of 21,000 years, which is the real interval for seasonal change. As a result, the northern and southern hemispheres flip their relation to perihelion and aphelion every 10,500 years rather than every 13,000, the difference revealing a lovely interplay among Earth, Sun, Moon, and planets.

Over the course of 21,000 years, the date of perihelion moves around the whole calendar of 365 days, becoming later and later each year at a rather amazing rate of one day every 21,000/365 = 58 years, less than a typical human lifetime. Celestial changes are not so slow after all.

Message from Mercury

All planets have changing perihelia, some advancing, others regressing, all caused by mutual gravitational interactions. While such motions may appear only of academic interest, one played a powerful role in the way we live now. By the end of the nineteenth century, astronomers had calculated the advance of Mercury's perihelion on the basis of Newton's laws to be around 530 seconds of arc per century. Mercury, however, knew better, the swift planet's perihelion clipping along at what is now measured at 574 seconds of arc per century. The finally determined shortfall of 43 seconds of arc per century drove everyone nuts. Some astronomers even fiddled with the inverse square part of Newton's law of gravity.

Then came Albert Einstein. In 1905 he published his theory of special relativity, which contained that most famous equation, $E = mc^2$, which lies at the core of solar power. A decade later, he published the General Theory, in which he postulated that gravity was a consequence of the bending of spacetime—his amalgam of space and time—by mass (for reasons nobody yet understands). Two powerful observations originally supported the theory. First: if the Sun bends spacetime, starlight passing close to it should be bent inward, making the stars near the Sun appear to be pushed outward (by a maximum of 1.75 seconds of arc) from its center. Photographs taken during solar eclipses show that is exactly what happens. Second, since gravity is a form of energy, and since energy and mass are strictly related, gravity has a mass equivalent. But since mass in turn "has gravity," gravity itself must have its own (albeit small) gravity, which changes our calculations of gravitational fields through the Solar System. When that is taken into account, we exactly recover Mercury's missing 43 seconds of arc.

The obscure motion of this little planet led the way. Without the theory, without taking relativistic distortions of space and time into account, we would not be able to hurl spacecraft accurately through the cosmos, nor, among other things, would the global positioning system work. Your car might then wind up in Keokuk rather than Kalamazoo. Is astronomy a "pure science" without application? Apparently not. Are we touched by the heavens? Apparently so, and in more ways than we might at first imagine.

Tip and Bend

Advance of perihelion and precession are only the beginning. Two more variations beg for attention. First is the tilt of the Earth's axis, the obliquity of the ecliptic. It's not constant, but instead is currently decreasing at the rate of 0.47 seconds of arc per year. If this movement sounds inconsequential, think for a moment of the implications. The latitudes of the tropics of Cancer and Capricorn must always be equivalent to the obliquity. A degree of latitude corresponds to 111 kilometers (69 miles) on the Earth's surface. Consequently, the two tropics are approaching the equator at a rate of 14 meters (16 yards) per year. That's 4 centimeters (1.5 inches) per day! Over the whole Earth, the tropics are shrinking at a rate of 1000 square kilometers, about the size of a small town, per year. And there is not a thing that can be done about it in spite of "Save the Tropics" bumper stickers. Few understand the matter better than the residents of Taiwan. In 1908 the government erected a huge monument to mark the Tropic of Cancer. It's now nearly 1.5 kilometers to the north of the tropical circle.

At the other ends of the world, the latitudes of the arctic and antarctic circles are the complements (90 minus) of the obliquity. As the tilt shrinks, they move at the same rate toward their respective poles. Not only are the tropics shrinking, but so are the arctic and antarctic, at a rate of nearly 500 square kilometers per year.

The angle that the Sun makes at noon with the Earth's surface determines the heating rate of the ground, and over the year is almost entirely responsible for the seasons. As the obliquity decreases, any given temperate latitude will get less heat in summer and must thereby cool. However, since the Sun will not get so low in winter, that season must become slightly warmer, and seasonal variations become less. If the tilt went all the way to zero, those in mid latitudes would feel a constant chill of early spring or fall, and the seasonal cycle would quit altogether.

The effect, however, is cyclic. The main cycle causes the tilt to vary between extremes of about 22 and 24.4 degrees over a 41,000-year period that is modulated by a longer million-year variance. The effect, while not large in itself, is noticeable. Recall the difference that 2.5 degrees of latitude (280 kilometers, 170 miles) in the northern United States can make in the coming of spring, the timing of fall's peak colors (figure about a week's difference per degree), and on summer's heat.

The planets are once again responsible. Because they have their own *orbital* tilts, they will usually be found above or below the Earth's orbital plane, and can therefore pull up or down on us. The result is another precession, this time not of the Earth's rotational axis, but of its *orbital* axis. This planetary precession acts in the direction opposite to the lunar-solar precession, countering it by moving the equinox counterclockwise, to the east, by 0.12 seconds per year. Since the Earth's orbital plane rotates against the rotational plane, our axial tilt changes. The two precessions together give the main 41,000-year tilt cycle.

Second, as the planets pull on the Earth, they slowly change its orbital eccentricity over a cycle a bit over 100,000 years long. The eccentricity determines how close the Earth is to the Sun at perihelion, and how far away it is at aphelion, and thus determines the degree of the modulation of the seasons (that is in turn modulated by the precession of the equinoxes). At the low end, the eccentricity can go nearly to zero, for all practical purposes eliminating perihelion and aphelion altogether. Currently at 1.7 percent, the eccentricity can go as high as 5.5 percent. At the upper end, the difference in solar heating at perihelion and aphelion becomes seriously large, up to 23

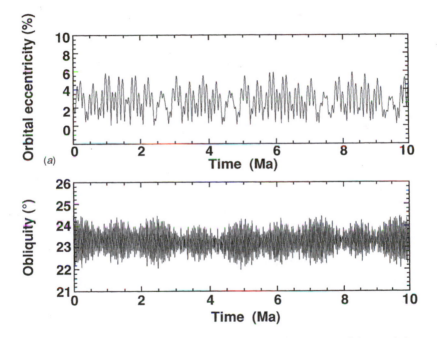

Figure 4.6 Over the past 10 million years (Ma), the eccentricity of the Earth has wobbled from near zero to almost 6 percent, while the obliquity of the ecliptic (in degrees) has shifted between 22 and nearly 24 1/2 degrees. Combination of these cycles with precession (plus perhaps others), all the result of outside gravitational influences, quite likely produces ice ages. From Linda A. Hinnov in *A Geologic Time Scale 2004*, New York: Cambridge University Press, 2004, reprinted with the permission of Cambridge University Press.

percent as opposed to the current 7 percent. Try *that* when perihelion coincides with summer and aphelion with winter!

Mister Milankovitch

The idea that changes in the Earth's positioning relative to the Sun might produce climate variations—and in the extreme, ice ages—goes back to the middle of the nineteenth century. Ideas, however, are by themselves insufficient. They need development, which was provided in the early twentieth century by the Serbian mathematician and scientist Milutin Milankovitch (1879–1958). He combined the three cycles (those of the tilt, precession, and eccentricity) and calculated how the sum of them would affect the heating rate at high latitudes. In his honor, the trio has become known as the "Milankovitch Cycles."

The resulting curve displays "rapid" oscillations that correspond to the precessional cycle, upon which are superimposed the longer-term variations caused by the tilt and eccentricity effects. Every 100,000 or so years, the amplitude (the peak-to-trough variation) falls to a minimum or rises to a maximum, the long interval corresponding to that of the eccentricity variation: and of the ice ages. Smaller variations seem to be linked to the other cycles.

Whether there is a genuine cause-and-effect is not clear. Just because two things happen together does not mean one causes the other (a true tale reserved for a later chapter). Such assumptions can create notably bad science. Though the peak-to-peak seasonal differences caused by changes in eccentricity are large, the annualized average difference in the heating rate between maximum and minimum eccentricity is not, just a small fraction of a percent, which seems insufficient. And cycles that should be present in the ice are not, in particular a long 400,000-year eccentricity term. Whatever the details, one idea is that changes in the Earth's orbit and orientation provide a collective trigger that kicks in some kind of positive feedback mechanism, which then drives the ice. (In positive feedback, variations in "A" cause "B" to change, which in turn produces increased changes in "A" that then produce increased changes in "B" and so on.) Such mechanisms may involve Earth's reflectivity, ocean currents, atmospheric carbon dioxide, and a variety of others.

We may also not be at the end of the orbital story. There is yet another, perhaps more powerful, cycle, one not considered by Milankovitch. The tilt of the Earth's orbit relative to the average plane of the Solar System (which is pretty much defined by Jupiter) varies over a 100,000-year period as well, from near zero to as high as 3 degrees. Our Solar System's disk is filled with dust from asteroid collisions and comet disintegration. The dust swept up by the Earth in turn may affect the heating rate on the ground, either increasing or decreasing it, depending on its composition. Cloud cover and rainfall could easily be affected. As we move in and out of the average plane, the amount of dust we accumulate changes. If the Earth lingers in the central plane, it will accrete more than when it simply passes through the central plane a couple times a year. The idea is loosely related to disasters that might be precipitated by the Sun's moving in and out of the plane of the Galaxy (again a subject for a later chapter). Perhaps a combination of all effects is responsible, each to different degrees. Though no one yet really knows, it's still widely accepted that orbit/orientation effects are ultimately at the heart of it all.

It could be much worse. While the Moon is the prime instigator of precession, at the same time it stabilizes the obliquity, keeping it within a very small range. Given no Moon, calculations suggest that the tilt would randomly change by tens of degrees, producing extreme seasonal changes over only a few million years, and making life on Earth nearly—if not truly—impossible. The effects of the ice ages would be small by comparison. Mars seems to suffer from such variation. Add that to the tides as another life-giving effect of our lunar companion. While lunar/planetary gravitational effects may occasionally wreak havoc, we would likely not be here without them.

For millennia, people have believed in various forms of astrology, that we and the Earth are affected by the planets and their changing configurations. Indeed we are, but not through magic. The means instead are through the simple physics of mutual gravity, as the Sun, Moon, and planets gently tug us back and forth in a never-ending orbital dance. Other effects, as we will see in the next pair of chapters, are not so gentle.

Chapter 5

The Accidental Asteroid

Bad science fiction stories, movies, and video games have brave astronauts entering the deadly asteroid belt between Mars and Jupiter where, with dazzling moves and a variety of weapons, the interplanetary travelers fend off a constant attack on their spaceship by huge, speeding boulders.

There indeed *are* a lot of asteroids—thousands, millions, billions (depending on where you want to draw the lower size limit). There is also an awful lot of room. Even given a billion of them, each would have a box over 100,000 kilometers across to roam around in. If there were but two cars in the world, what would be the odds of them crashing into each other? Collisions between asteroids and spaceships are thus highly unlikely. As evidence, we've sent half a dozen real spacecraft through the belt (the two *Pioneers*, the twin *Voyagers*, the Jupiter-orbiting *Galileo*, and the Saturn-orbiting *Cassini*) with nary a ding.

Yet crash they do: but into each other. If you think of an asteroid as its own unprotected spaceship, it will indeed collide with other asteroids, not just once, but countless numbers of times, leaving each one beaten, battered, shaken, and covered with dust. The difference between our human spacecraft and a real asteroid is that the asteroids are not just passing through the belt, but are stuck there for billions of years. All they need for disaster is time. Those that *do* escape have the propensity for colliding with the planets, including our own, and therein lies the real story, which is no longer science fiction.

Numerology

The story of the asteroids goes back to 1766 and Johann Titius (1729–1796), whose "discovery" of a sequencing in planetary distances was usurped by the eminent German astronomer Johannes Bode (1747–1826). Sleight of hand, a "card trick," if you will. Lay out a string of number 4s. Below each, from left to right, place a sequence of numbers that starts with 0, goes to 3, and then doubles:

4	4	4	4	4	4	4	4	4	4
0	3	6	12	24	48	96	192	384	768

Add them up and divide by 10 to get

0.4	0.7	1.0	1.6	2.8	5.0	10.0	19.6	39.8	77.2

Now write down the distances (in Astronomical Units, AU) from the Sun of the planets known at the time, matching as best you can:

0.4	0.7	1.0	1.5	...	5.2	9.5	19.2
Mercury	Earth		woops		Saturn		
	Venus	Mars		Jupiter	Uranus		

The fit, and what is now known as "Bode's Law," is quite amazing. Or not. Since the planets lie outward from the Sun in some sort of geometrical progression that has to do with formation and mutual gravity, it would be surprising *not* to find some kind of numeralogic "law."

Given the amazing match between prediction and reality, however, there seemingly *must* be an undiscovered body at or near 2.8 AU. So astronomers looked for it. And on the birthday of the nineteenth century, January 1, 1801, they—actually Giuseppe Piazzi—found it. Right at (as calculated later by the brilliant mathematician Karl Friedrich Gauss) 2.8 AU. Piazzi named it Ceres, for the Roman goddess of grain.

A new planet! But, far fainter than Mars, not even visible to the naked eye, could it really be such? (The controversy over Pluto—look to next chapter—is hardly a new kind of issue.) Apparently not, since Heinrich Olbers found another (Pallas) at the same dis-

Figure 5.1. Stony asteroid 243 Ida orbits within the main belt 2.86 Astronomical Units from the Sun, taking 4.8 years to make a full circuit. Though large, 56 kilometers (35 miles) long, it was broken down from an even bigger rock in a collision that left it shattered. Additional cratering impacts have added to the insult. The little chip to the right is its satellite, Dactyl. *Galileo* image, NASA.

tance just a year later. Then came Juno (1804), and Vesta (1807, the only one visible without telescopic aid). By 1872 the count was 100. The number slowly crept upward through the twentieth century into the tens of thousands. Then with new technologies and dedicated telescopes, the dam burst to the point that no book can keep up with them, hundreds of thousands known from discovery rates of thousands per month.

And they all have names. The biggest and oldest known were called after mythological characters, and when these ran out they were named after the famous, the not-so-famous, the infamous, friends, family, dogs, whatever worked. The discovery order usually precedes the name (1 Ceres, 2 Juno, etc.). Now most of the "names" are just numbers in order of discovery, the exceptions being those individuals who are being honored.

While the leader, Ceres, is the biggest, it is still not very large, giving rise to the collective names of "asteroid" (the erroneous "little star"), "planetoid" (not popular), and "minor planet." Ceres is only 934 kilometers (580 miles) in diameter, a mere quarter the size of the Moon, which in turn is a quarter the size of Earth. Pallas and Vesta are just half Ceres's size, while Juno is down to a quarter of Ceres. Between Vesta and eighth-ranked Juno are four others. Only 25 are known to be more than 200 kilometers across. Most are just big rocks. Still, there may be a million or more larger than 100 kilometers wide.

From their huge numbers, one might think that the asteroids would be gravitationally important to the rest of the Solar System, perhaps deflecting planets and changing orbits. Not by very much. Ceres has a mass of but a ten-thousandth that of Earth, while the total mass of ALL of them (estimated from the run of number with diameter) is only three times *that*.

The vast majority is bunched in a "main belt" between about 1.8 and 3.5 Astronomical Units from the Sun (from 1.2 times the size of the Martian orbit to 0.7 times that of Jupiter's) that averages out at the distance of Ceres, 2.8 AU. Some, however, stray far outside the main belt, where they range from the orbit of Jupiter to within the orbit of Mars. Even to within the orbit of the Earth.

(While Bode's law certainly was successful in leading us to the asteroids, it breaks down beyond Uranus. The "prediction" for the next planet is 39.8 AU, while Neptune falls at 30.1 AU. Pluto, over which planethood is strongly argued, "should be" at 77.5 AU, but is at 39.5, where Neptune ought to be. Moreover, Neptune was apparently not born in its current position, but more where Uranus is now. Given that there is no physical basis for the "law," such failure is not surprising.)

Origins

How did they all get there? The asteroids (and the comets of the next chapter) seem to be natural by-products of planetary formation. The spaces between the stars are filled with a lumpy amalgam of gas and dust. The mix of chemical elements is about the same as we find in the Sun, with hydrogen dominating (at the 90 percent level, helium next at near 10 percent). The dust is a complex mixture of silicates and carbon grains mixed with metals, all of which are the ejecta of dying stars.

The Sun was born when a lump of cold dusty gas buried within a thick cloud made mostly of hydrogen molecules had enough self-gravity to contract. As our star-to-be squeezed down, it rotated faster (conservation of angular momentum in action). The stuff that did not fall into it flattened to become a dirty whirling circumstellar disk. While the new Sun stabilized to fuse its internal hydrogen into helium, the dust grains within the disk began to collide and stick, new larger grains growing from smaller ones. Friction with the gas in the disk moved the particles inward to keep the collisions going, further consolidating the grains. At some point, they would have become visible to the eye, and then they grew to millimeter size, centimeter, and larger to meter and kilometer, so large that the gravity of the bigger bodies—now called "planetesimals" for "infinitesimal planets"—could start attracting and then swallowing the smaller pieces, their growth now unrestrained.

Free atoms and molecules from the gas could also condense onto the grains and planetesimals, adding to their mass. But here we encounter a sort of "continental divide." In the inner Solar System, where temperatures were relatively high as a result of solar heat and high viscosity, only atoms and molecular compounds with high condensation temperatures—metals, silicates—could attach themselves. It was too warm where we now live for one of the most common of all compounds, water (in gaseous form), to be incorporated. However, at an orbital radius poetically called the "snowline" (around 5 AU), solar heat dropped enough for the water vapor (and a variety of other volatile chemicals) to condense onto grains and larger bodies as ices.

At the end of the creative process, some 10 to 100 million years after the Sun was born, most of the planetesimals had consolidated into but eight planets or planetary cores. In the inner system, the four winners—Mercury through Mars—stayed that way. They came out small from lack of condensing matter, and very dry. There were at first most likely more such bodies, but the Earth seems to have taken one out in a giant collision whose debris formed our Moon, which gave us tides, controlled and stabilized our rotation and axial tilt, and in good part allowed the development of life. Venus's backward rotation suggests its own gigantic collision.

Not only did the outer planetesimals have an abundant ice supply, but the resulting larger planetary cores could also attract and retain the highly volatile but super-abundant gaseous hydrogen and helium, allowing them to grow huge. Jupiter and Saturn had the best possibilities, with sufficiently low temperatures and maximized disk densities. Farther out, though, the amount of disk matter dropped off, and Uranus and Neptune turned out smaller and denser. Beyond Neptune, the colliding iceballs stopped at the small planetesimal stage, where we now see a rich belt of small chunks that grew large enough to make Pluto and a few others, but no farther. Collisions among the larger bodies then broke many of them back down, their host ultimately creating one kingdom of comets. Within the first billion or so years, the newly formed planets violently swept up much of the remaining debris, a lot of which was being tossed around by planets that were still trying to find their final orbits. The collisions resulted in the heavy cratering seen on the Moon and inner planets and on the icy surfaces of the outer satellites (erosive processes wiping Earth fairly clean).

Had there been no giant planet—Jupiter—the primitive original bodies of what we now call the asteroid belt would probably have consolidated into a real planet. But Jupiter's powerful gravity disturbed the orbits, forcing the small bodies to cross each others' paths at high velocities. As a result, after being allowed to grow to a certain size, they too began to collide, breaking themselves apart. Over the aeons, they have ground themselves back nearly to dust, with only a few large ones surviving the deadly mutual onslaught. Jupiter then stirred the asteroids into the thick disk that they occupy today.

We've been out there to look. On its way to Jupiter, *Galileo* took close-up images of asteroids 941 Gaspra and 243 Ida. Both are highly irregular fractured bodies (Gaspra 19 × 11 kilometers across, Ida similar but nearly three times as large) with broken ends and filled with craters showing what 5 billion years of collisions can do. Ida even has a small satellite that was probably cracked off its surface. Even better, we've landed on a pair. The *NEAR* (Near

Figure 5.2. The Moon, here in its waxing gibbous phase, is covered with craters from ancient impacts. The circular features, the "maria," are somewhat younger lava-filled impact basins. Collisions continue today, though at a vastly lower rate. While the Earth must have been similarly blasted, only the more recent craters have survived the effects of erosion and other means of destruction. The Moon, with no air or water, has kept much of the full record. Mark Killion.

Earth Asteroid Rendezvous) mission not only orbited 433 Eros, but landed a camera that took close-ups on its descent and kept transmitting data for weeks. Though plastered with craters from countless collisions, the 16-kilometer-wide body is also covered with a deep fine dust created by the impacts. The Japanese *Hayabusa* craft also not only took images of tiny Itokawa (which comes close to Earth), but briefly touched down in a failed attempt to collect samples and bring them back. Some asteroids are not even "solid." After being totally smashed, they stitched themselves back together under their weak gravity to make "rubble piles."

Jupiter's disturbances similarly seem to have prevented Mars from growing terribly large, keeping it smaller than Earth. Having a happy combination of being farther from the Sun than Mercury or Venus (where there was more available raw material) and farther from Jupiter than Mars, our planet turned out the largest of the inner four. Closest to the bulk of asteroids, the red planet is hammered by the stray asteroids that hang outside the main belt. Its two tiny moons may in fact be asteroidal debris. Earth gets it too. But first we have to bring them here.

Child on a Swing

Go to the park. Put your youngster or friend on a swing. Pull back and let the swing—swing. It has a natural period that depends on the length of the chains that hold it. (And on the gravity of the planet. Given the same swing, our Martian's kid would swing more slowly, but could also go a lot higher.) If you leave him or her swinging, the swing slows down; so to keep your friend happy, you push. But not at random, as that would really annoy the swinger. If you push at the natural period, you not only keep the swing going, but increase the size—the "amplitude"—of the swing's arc. Pushing at double the period works well, too, as does triple, though once you get much beyond that, the effectiveness wears off. If you get somebody on the other side, you can alternate and push with half the period as well. In all cases you are "resonating" with the swing. Nature produces its own resonances. And oh the effects!

The most famous and obvious resonance involves the rings of Saturn. The rings are made of countless small icy rocks each in its own orbit about the planet, possibly the debris of some smashed up satellite. The rings shepherd themselves through constant collisions to create a bright disk nearly 300,000 kilometers in diameter (double Saturn's size, which itself is nine time's Earth's diameter), but well under a kilometer thick. About 85 percent of the way out is a dark gap, the "Cassini division," named after the seventeenth-century astronomer who realized that Galileo's "jug handles" were actually rings. They can be seen even in a small backyard telescope.

Kepler's laws apply anywhere. For either Saturn's ring particles or satellites, the squares of the orbital periods are proportional to the cubes of the orbital radii: double the distance, and the travel time goes up by a factor of 2.8. Put a particle at the distance of the Cassini division from Saturn's center and it will have a period exactly half that of one of Saturn's more substantial moons, Mimas. Every time Mimas makes one orbit, our particle makes two. If they are lined up at any given moment, then one Mimas orbit later, they are lined up again. Gravitational effects build up. Mimas becomes the pusher, the little particle the kid on the swing. Not only does the particle swing away more from its given orbit, it gets kicked out of the orbit altogether and into another one so as to get away from its irritating big brother. The Cassini division then naturally empties of any particles that might move into it through collision or other means.

Such resonances abound in the planetary system, and can both eject and trap. Jupiter's big four satellites (from inside out Io, Europa, Ganymede, and Callisto) were discovered by Galileo in 1609. They are so large and bright—comparable to the Moon and Mercury—that they can be seen with binoculars. The inner trio is locked in resonant embrace. Every time Ganymede orbits once, Europa goes around twice, and Io four times. Thus they are constantly lining up. They are each too big to be ejected by the other. Instead, Ganymede and Europa gang up on smaller Io to push it around a bit from its average circular orbit.

A slight change in the distance of the Moon from Earth produces a large change in ocean tides. Tides raised by Jupiter on little Io are far fiercer than any effects the Moon and Earth can have on each other. Resonant changes in Io's distance cause huge tidal changes. As a result, Io is continuously flexed. Go to the grocery. Squeeze an orange to test it for freshness. Now squeeze a million

Figure 5.3. The brightest two of Saturn's multiple rings are separated by the dark Cassini division, the result of a disturbing resonance with the satellite Mimas that ejects ring particles. The planet's clouds are made of ammonia crystals. The bright patch is a storm system. Saturn spins so fast, in just 10 hours and 40 minutes, and is made of such light stuff (mostly hydrogen), that it bulges noticeably at the equator. Hubble Space Telescope: Reta Beebe (New Mexico State University), D. Gilmore, L. Bergeron (STScI), and NASA.

times per second. The massive flexing heats the inside, and you are sprayed with hot orange juice bursting through the skin. Io is similarly heated by tidal flexing, and out squirts hot sulfur-laden silicate rock to make the satellite the most volcanic body known. Europa is tidally shifted a bit too, which many think heats a sub-surface ice layer into a warm, though ice-coated, ocean. One that we speculate might be a place for life. All because of resonances.

Resonating Asteroids

Accumulate enough data, and all sorts of patterns begin to emerge. Each asteroid has its own elliptical path with all the usual numbers associated with it, including the orbital period in years (P), the orbital size (a, the semimajor axis of the ellipse, equal to the average distance between the asteroid and the Sun) in AU, and the eccentricity (how much the orbit deviates from a circle). A plot of the orbital sizes of the known asteroids reveals a set of prominent gaps first identified by Daniel Kirkwood in 1866.

There are almost no asteroids with average distances that lie in rather broad avenues centered on 2.06, 2.50, and 3.28 AU. How do they know not to assemble there? Once again invoke Kepler and his remarkable third law, P-squared proportional to a-cubed. Do the math to get the periods that the asteroids in the Kirkwood Gaps must have, and respectively find 2.96, 3.95, and 5.92 years, which are one-fourth, one-third, and one-half that of Jupiter's 11.86! Like the Cassini division in Saturn's rings and Io's tidal heating, the Kirkwood gaps are wonderful examples of gravitational resonances.

Actual positional maps of the asteroids reveal nothing. The orbital eccentricities keep the asteroids with non-resonating periods moving in and out of the gaps such that the "voids" are hidden from view. But let the asteroid try to take on a critical average distance that allows the period to be in a simple ratio to that of Jupiter's, and wham! Out it goes. The 2:1 (the asteroid orbiting twice for every Jovian turn) and 4:1 gaps effectively cut the main belt down to between 2 and 3.2 AU. The Sun may control the planetary system, but Jupiter is clearly the "underboss" in charge of the asteroids—and anything else that gets too close.

Other simple ratios have similar, though less prominent, effects, notably the 7:2 ratio (7 asteroid turns for every 2 of Jupiter) at 2.26 AU, the 5:2 at 2.82, and the 7:3 at 2.96 AU. Beyond 3.5 AU, however, outside the main belt where few asteroids roam, we see the reverse effect. At the 3:2 ratio (3.97 AU), Jupiter acts oppositely and instead imprisons a substantial number. At the 4:3 (4.29 AU) is a shepherded smattering. And then, of all places, at 1:1, right in Jupiter's orbit itself, are two gangs of them called the "Trojans," one averaging 60 degrees ahead (in a very large stable gravitational "well"), the other following behind by the same angle.

Kick on Goal

Asteroid collisions continue to this day. We see several "families" of them with similar orbits and reflection characteristics that make it obvious that they are the collected remains of smashed-up parent bodies. A collision can also send debris flying out in all directions, where a chunk might enter a resonating orbit and be strongly rejected by Jupiter. An ordinary planetary perturbation can do the same thing. The offending asteroid, which could be anything from a peanut to a multi-kilometer-wide microplanet, could easily be thrown into a new orbit in which it crosses that of the Earth.

And we see them. A whole family of otherwise unrelated asteroids, the "Amors," are named after 1221 Amor. Amor is a kilometer-wide lump of rock with an orbital period of 2.6 years and an average solar distance of 1.92 AU. While it gets close, Amor—and the rest of its family—stays *outside* Earth's orbital door. Then there are the "Apollos," whose namesake (1862 Apollo) is half again as large as Amor and goes around the Sun in 1.8 years. With a semimajor axis of 1.47 AU, it—and the rest of the Apollos—open the door and, because of their orbital eccentricities, come *inside* Earth's orbit. Finally, there are the "Atens," after 2062 Aten (1.1 kilometer wide), whose orbit averages just under an AU, its period just under a year (347 days). The Atens, lying entirely inside our path, have opened the door, gone through, and slammed it behind them.

Can the goal itself be scored? Planetary influences constantly alter asteroid orbits. We are a small target. But the possibility is there for a strike. Not for nothing does this set of three groups provide us with "Potentially Hazardous Asteroids." They are technically defined as any body more than 150 meters across that can get within 0.05 Astronomical Units of Earth, though you would not want anything of any real size hitting you. However, possibility, even probability, are not actuality. We need evidence. And there is no lack of it.

A Shocking Story

Go outside for a breath of cool night air. Suddenly, a flaming thunderbolt flashes across the sky as if thrown by Zeus himself. Bright enough to light up the surroundings, trailed by shed bits and pieces, it explodes, its radiant shards falling toward the horizon. You have just seen a *meteor* (from Greek, a "thing in the air") of a special kind, a *fireball*, a chunk of celestial debris that is passing through the Earth's atmosphere. It's so bright that you think its remnants must have landed just a mile or two away. Looking, you find nothing, no glowing coals. In the paper the next day, you read that a town hundreds of miles distant was bombarded by a shower of stones from the sky: the meteor seems to have spawned meteor*ites*—rocks that hit us. Meteors are remarkably common, most by far produced by the rocky stuff flaked off comets under the action of sunlight (a tale for the next chapter). Those that give us meteorites are vastly more rare and have another origin.

The speed of the incoming body (the "meteoroid") is huge. The Earth orbits the Sun at nearly 30 kilometers per second. Depending on direction and angle and pre-strike orbit, the invader could easily be going 50 kilometers per second or even more (in freeway terms, 112,000 miles per hour). At that rate, the thing could go across the continental United States in a minute and a half! There is considerable variation, however. In the evening, we are facing away from the direction of the Earth's orbital motion, looking toward the back of the train, so to speak. The meteoroids have to catch up with us, are therefore generally moving more slowly than average, and appear reddish and more lazy as they cross the sky. In the morning hours we face the way the Earth is going, so our orbital speed is added to that of a space rock. The result is a hot, blue-white, and very fast bullet plowing through the upper air. Whatever the actual speed, the energy release upon hitting us is spectacular, and the meteoroid does not have to be very big to be very bright. Most ordinary meteors are not much bigger than garden peas, even grains of sand. Big ones, the fireballs, may typically be only a meter or so across.

When any body moves through a fluid at a speed faster than that of the natural wave motion, the fluid's atoms or molecules cannot get out of the way fast enough, and pile up into a "shock wave." The most familiar example is the wake that comes off the bow of a speeding power boat (rendering it a "bow shock"), which can be so strong it rocks the dock and surrounding vessels. It can swamp a rowboat. Another familiar example is the bow shock that trails from the fuselage of a jet plane flying faster than sound (sound being a compression wave in the air that moves at 1200 kilometers, 750 miles, per hour). When the shock passes over you, you hear a loud "thud," the "sonic boom." (Sonic booms are commonly misconceived to occur when the plane "breaks the sound barrier." Instead, the "boom" is produced continuously as long as the craft is supersonic.)

We need not even have to have speeding "bodies." Out in space, the solar wind sets up shock waves when it encounters planetary magnetic fields, including that of Earth, to create aurorae. At a distance of some 100 Astronomical Units from the Sun, it stalls against the gases of interstellar space, creating a shocked bubble—the "heliosphere"—within which we orbit and live. Shocks from exploding stars accelerate chapter 7's cosmic ray particles that slam into the Earth at nearly the speed of light; such shocks also compress the interstellar gases to initiate star formation. Without shock waves, we would not exist.

When a meteoroid first encounters a thickening atmosphere at an altitude of 100 kilometers or so (50–75 miles), it is moving vastly faster than the speed of sound, in jet airplane terms at perhaps Mach 50 (Mach 1 equaling the speed of sound), 20 times the speed of the fastest military craft, half a dozen times the re-entry velocity of an orbiting spacecraft. It thereby sets up a violent shock that heats the air through which it passes to around 6000 Celsius—10,000 F—as hot as the surface of the Sun. What you see as the meteor is not so much the heated body, but the glow from the ionized tube of air through which the meteoroid passes ("ions" being atoms and molecules that have lost electrons and have become "electrified"). It takes time for the air to cool and neutralize, so the glow lingers after the fireball has passed, sometimes for several minutes.

You might even *hear* it. Becoming tangled in the hot, electrified tube, the Earth's magnetic field lines generate very low frequency radio waves. The waves immediately set up magnetic and electrical disturbances in the ground, resulting in "electrophonic sounds" that can sometimes be heard during the body's passage overhead. About 5 minutes later you might also hear the dull thud of the sonic boom itself.

Though the surface of the meteoroid will suffer melting, if the body is large and strong enough, it hits as the meteorite. Perhaps it will even find its way into a collection. If really large it might destroy itself—and everything around it—on impact. Maybe even globally.

Heaven's Rocks

Meteorites tell their own stories, even though many of their characteristics are still hard to understand. Unlike any Earth rock (you can't fool an expert), they come in two broad varieties, "stones" and "irons." The irons are the most common in museum collections. They resist destruction through weathering, and also truly stand out, so are the easiest to find on or in the ground. When we actually look at the numbers that fall, however, the irons are by far the rarer, constituting only about 5 percent of the total population. The stones make nearly all the rest. In between there is a tiny fraction of "stony-irons" that are a sort of blend of the two.

Iron meteorites are one of the few forms of "free iron," iron not chemically mixed with other elements to make an iron ore as usually found on Earth. Not entirely pure, they are typically a mix of 80–95 percent iron with the rest nickel (two of the more abundant of metals). In the hands of a metalsmith, irons can be used to make quite beautiful knives and other implements, some—if the nickel content is right—nearly impervious to rust. When sliced, polished, and etched with acid, they are striking. On the shiny surface is a network of fine lines that reveal huge metallic crystals (the "Widmanstätten pattern" after the nineteenth-century Viennese Count Alois von Widmanstätten, who discovered the procedure). Creation of such large crystals takes millions of years of cooling from a once-molten state. Iron meteorites can't be faked.

The stones have their own unique characteristics. Nearly 90 percent of them are made of tiny round balls called "chondrules" stitched together in a silicate rock matrix. When the meteorite is sliced open, the chondrules appear as round circles, one next to the other next to the other. They bear all the features of being

Figure 5.4. The oldest thing you could ever hold, this meteorite—a carbonaceous chondrite 4.6 billion years old—came from a shower of stones that hit the town of Allende, Mexico, in 1969. The origins of the small round chondrules are unknown. J. B. Kaler.

flash-heated and suddenly cooled, then later bonded together. A fraction of chondritic stones is loaded with carbon. These "carbonaceous chondrites" are rather fragile, and contain small amounts of water, tiny diamonds, and even organic compounds. The remainder of the stones have no chondrules, but still have a chemistry quite different from Earth rocks.

Iron meteoroids are strong, and can easily survive atmospheric passage. The stones, however, are often fragile, and are easily disrupted by heating and the violent shocks. Stony fireballs therefore often break apart into several—even hundreds—of fragments, causing rains of rocks on the ground. Large irons are thus relatively common. Found stones tend to be small.

Interpretation

These "rocks"—going from iron to stony-iron to stone—seem to form some sort of structural continuum. Except for volatile hydrogen and helium and a few other exceptions, stony meteorites have chemical compositions rather similar to the Sun. Indeed, where the solar element is unobserved, meteorites become the solar proxy.

Among the more common radioactive elements is uranium, whose most common isotope decays into a particular isotope of lead. Measurement of the amount of the product (lead-206) to that of the producer (uranium-238), along with the known decay rate and allowance for the lead-206 initially present, yields the age of the rock since the "clock" was set by solidification. A number of other similar ratios (thorium-lead, potassium-argon) work just as well, if not better. Meteorites turn out to be the oldest things known, the carbonaceous chondrites coming in at 4.6 billion years of age. They appear to be original to the Solar System.

The Earth's average density of well over 5 grams per cubic centimeter is nearly double that of surface rock. There must be something very dense inside, a metal, to compensate. Analysis of earthquake waves (which penetrate the entire planet) reveal a distinctive core about halfway in. Common iron and nickel have the right densities and abundances to give the measured overall terrestrial density. The other inner planets have nickel-iron cores too, Mercury's huge, Mars's small. The Earth, born through violent collision with impacting planetesimals, even with small planets, was heated to or near the melting point. From its mix of solar-like chemical elements, the heavy iron and nickel sank to the inside to create the core, while the lighter silicate rock floated to the top to make the thick outer mantle, our planet—and others—naturally "differentiating."

We thus explain at least some kinds of meteorite as the remnants of the asteroids that had differentiated before suffering massive collisions. The irons are from the interior cores (giving credence to the Earth having an iron core), the stones from the outer mantles, the stony-irons from the interfaces. Undifferentiated rocky asteroids add to the mix. The chondrites, particularly the carbonaceous ones, never developed very much and are the remnants of perhaps the original small planetesimals that together would have created a planet—had Jupiter not gotten in the way. That the chemistry of primitive stones matches that of the Sun fits right in. Meteorites are asteroids that have strayed inward—thanks to Jupiter—and hit the Earth.

Powerful further evidence is provided by orbits. On a precious few occasions (five, actually), the fireball parent of a meteorite was tracked such that its path could be triangulated and the orbit of the incoming meteoroid calculated. The bodies trace themselves right back into the asteroid belt. Kick on goal completed, and we come full circle. While Zeus throws mythical thunderbolts, there is nothing mythlike about the actions of his Roman namesake. Meteorites are genuine asteroids thrown to us by Jupiter. Not only have we made the broad connection, but through their reflective properties we can match the different kinds of meteorites with different kinds of asteroids (stones, metals, etc.). In the best case, we are pretty sure that we have meteorites that are pieces broken from Pallas. Meteorites allow us to view alien "planets" close up and to bring them into the lab for study. Or at least a set of them, since Nature tosses us what she wants.

A Steady Rain

While thunderbolts can be amusing to see in the distance, they can be devastatingly destructive. So can asteroids cum meteorites. We live in a constant rain of them, 1000 or so per day heavier than an arbitrary 10-gram limit. However, rather like the asteroid belt itself, though the number of potential meteorites is large, so is space. Given the size of Earth, you would be lucky to have one fall within a few kilometers of you in your lifetime. And you'd likely never see it. You are such a tiny target that being struck by one is more than improbable. Don't buy meteorite insurance.

On the other hand, there are lots of people in the world, and at some point, somebody gets it, if only indirectly. The International Meteorite Collectors Association lists 78 strikes of "HAMS" ("humans, animals, and man-made objects") over the course of the twentieth century—nearly one a year. The list includes horses, cows, dogs, houses, cars, mailboxes, and yes, people, some of whom were actually injured, though nonfatally.

These are not simple rock-throwing incidents. The remarkable Peekskill (New York) fireball meteor, caught by several video cameras, was first seen over West Virginia. Zipping along at 15 kilometers per second, it broke up at an altitude of 45 kilometers, and became heaven's shotgun, spraying a huge area of New York State. Only one piece was actually found: the one that took dead aim on the trunk of a woman's parked car. Though braked by friction with the air, it still went through both the trunk lid and floor and then sat itself in the driveway. All worked out in the end, though, as the achondrite was worth the price of another auto. (The original car achieved some fame. An entrepreneur sawed off the back end and took it to rock and gem shows, where for a few dollars a visitor could have a picture taken with his or her arm stuck in the hole.) Then there was the stone that hit the house in New Orleans in 2003. It went through the roof, the first floor, and wound up beneath the basement, destroying things as it went. About once a year, so if you hear noises . . .

While none of us is ever likely to be struck by meteoric rain, we cannot escape the steady drizzle. Asteroid collisions create vast amounts of dust and small particles. Re-radiation of energy acquired through absorption of sunlight induces a drag on the dust grains that moves them sunward, while for larger particles the effect can be to move them either inward or outward. Eventually, they reach the Earth, where they filter downward to the surface. Breakup of meteoroids adds to this mix of "micrometeorites," as does cometary debris. It's all around you. Now smell not just the roses . . .

And Then the Hailstorm

The size distribution of meteorites before any destruction by the Earth's atmosphere reflects that of the asteroid parents. Little ones are amazingly common, big ones rare. But the big ones DO hit. With spectacular, even deadly, result. They also bring new life. We can even make a case that, were it not for such immense impacts, we—humanity—would not be here.

On the plains of northern Arizona lies a huge hole in the ground 1.2 kilometers (3/4 mile) across and 175 meters (570 feet) deep. From the air, it looks exactly like a crater on the Moon. The surrounding desert floor is littered with (or was, before collectors

Figure 5.5. Meteor (Barringer) Crater, a good fraction of a mile across, is the result of the impact of a huge iron asteroid perhaps 50,000 years ago. The road and structures at the bottom are from an attempt to mine the valuable meteoritic carcass, which instead mostly destroyed itself. J. B. Kaler.

swept them up) chunks of meteoric iron. The culprit was an iron meteorite that partially broke up prior to impact and mostly vaporized. Impact craters are caused when a massive body hits with enough energy to compress, fracture, even melt the rock of Earth. When the impacting pressure is relieved, the broken, partly molten, rock bounces back like a bedspring, which throws the shattered debris outward, leaving a cavity with a surrounding raised rim. The ejecta (sometimes recovered as glassy rocks called *tektites*) can travel for hundreds of kilometers. The energetics of Arizona's *Meteor Crater* suggests a parent impactor the size of a house weighing in at over 60,000 tons. Nothing is left but scraps. Dug some 50,000 years ago, there was no one around to witness, or be extinguished by, it.

Such craters cover Earth, about 170 known and confirmed. Most have been eroded, covered, or filled by wind-blown dust or glacial debris, but can still be identified by circular depressions, fractured surface or subsurface rock, or other geological clues. While few are as pristine as Meteor Crater, many are much larger. Germany's Rieskessel (the giant's kettle) is a shallow bowl 24 kilometers (15 miles) across. The crater's glassy tektites, called "moldavites" after the Moldau Valley, are found in the former Czechoslovakia. Clearwater Lakes in southern Quebec, a pair of craters each roughly 30 kilometers in diameter, were caused by either a double asteroid (a number are known) or one that broke in half on entry to the Earth's atmosphere when it struck around 36 million years ago. The Chesapeake Bay structure (which may be the result of a comet impact) is 90 kilometers across. Five are even larger. Back in Quebec is 100-kilometer-wide Manicouagan, a water-filled ring over 200 million years old. Barely visible, Ontario's Sudbury structure comes in at 250 kilometers (155 miles) across, while South Africa's Vredefort is 300 kilometers wide. Both are 2 billion years old.

One event in particular is strongly related to a mass extinction, and others seem possible as well.

Wipeout

Within our little slice of time on Earth, life seems pretty stable. The record of the past half-billion years, punctuated with mass extinctions of whole species, tells otherwise. Such events, which seem to average every 100 million years or so, commonly mark the boundaries of geologic ages. The most recent of them was the extinction of the dinosaurs 65 million years ago along with three-fourths of the Earth's other species, which defines the "K-T boundary," when the "Cretaceous" Period (which starts with a "K" in German) of the Mesozoic Era changed to the "Tertiary" Period of the Cenozoic. Similar events ended the Triassic 208 million years ago, the Permian at minus 245 million (with 95 percent of the species vanishing, the worst known), the Devonian (360 million), and the Ordovician (440 million). In between (and more recently than the K-T) were a number of "minor" extinctions (only minor if it is not YOUR species), including one at 35 million years ago (late Eocene) and the most recent at –11,000.

While ice ages play a role, as seen in the ending of the last one 11,000 years ago, they cannot account for the most massive disasters. We are slowly coming to see that they might be related to a variety of astronomical events, to the most immediately dramatic of "heaven's touches," including asteroid impacts. Compared to the chemical composition of the Earth's crust, meteorites are especially abundant in the element iridium (which, when the Earth formed, was carried downward into the core, making its crust iridium-deficient). Around the world is a layer of iridium-rich rock that is dated by radioactive methods to be 65 million years old—the time of the K-T boundary.

The iridium layer was the first clue that global destruction from a massive asteroid strike was responsible. Others abound. The layer also contains shocked quartz crystals that would have been produced via impact, tiny balls of glass, soot, and tektites scattered into what is now the western United States. The final evidence was the crater itself, Chicxulub, which cuts across the northern coast of the Yucatan peninsula in Mexico. It too is dated at 65 million years in the past. Though buried and invisible beneath sediments, deep drilling as well as gravity measures allow the tracing of its mighty curve. At 170 kilometers (105 miles) across, it is one of the largest terrestrial craters known.

The crater's size, and all the other evidence, suggests an impact by an asteroid some 10 kilometers—7 miles—wide: the depth of the deepest ocean trench. The heat from the blast seems to have caused massive forest fires in the Americas, Africa, and Asia. Dust, smoke, and aerosols raised by the impact would have been launched high into the atmosphere to be carried around the world, so much of the stuff that the Earth was plunged into a global winter night that cut off photosynthesis for months, even years. With no sunlight, no food supply, much of life just curled up and died.

Though the crater itself is invisible to the eye, its effects are most certainly not. Surrounding the land portion of the great pit is a ring of "cenotes," sinkholes carved in vulnerable limestone by water. Somehow, the impact seems to have structured the bedrock geology into a web of underground streams. The ancient Mayans (whose civilization flourished for centuries before being destroyed

Figure 5.6. In the twilight of the dinosaurs: Mexico's Chichen-Itza has one of many sinkholes—cenotes—created by the powerful asteroid strike that destroyed the ancient beasts. J. B. Kaler.

for mysterious reasons) used the water-filled pits as natural reservoirs. The most famed is the 60-meter-wide (200 feet) Sacred Cenote at that great center of Mayan culture, Chichen-Itza. The Mayans threw treasure, animals, and apparently some of their citizenry (or enemies) into the 35-meter-deep (115 feet) water as sacrifices to the rain god, much as the dinosaurs were sacrificed to the accidental asteroid. Ironically, they also built an observatory in homage to the sky (and its link to Earth), from which came the devastating event.

No other mass extinction has been so tightly correlated with an impact. But one other may come close. Gravity and subsurface radar maps of eastern Antarctica possibly reveal a monstrous crater buried under a mile of ice. Up to 500 kilometers (300 miles) wide, the structure is the largest known, dwarfing even Vredefort. The local geology suggests a rough age of about 250 million years, which more or less coincides with the Permian extinction. If the structure really was caused by impact (and the only way to know for sure is to look for shattered rock), it is at least possible—though very speculative—that most of Earth's life was taken out by a 50-kilometer-wide (30-mile) asteroid. Any of those in a similar range could have caused similar destruction. We can only speculate on what effects the "smaller" events, like those that produced Manic-ouagan, could have had.

But we have to watch what we mean by "disaster." The end of the Permian opened the way to the dinosaurs, while the void left at the K-T boundary left the world to mammals. If an accidental asteroid DID cause the dino-annihilation, or even was just the final coup de gras, we owe to it our safety from the giant beasts.

And Now?

One of the more common questions asked of an astronomer is "What's the chance of such a giant impact happening again?" Answer: "100 percent." It's just a matter of time. The lesson was vigorously learned in the summer of 1908, when a "modest" asteroid estimated to have been (at most) about 30 meters in diameter, weighing 40,000 tons, struck the Tunguska region of Siberia. Stories tell of extraordinary devastation, of seeing a sunlike fireball during the day, of hearing explosions as far as 500 kilometers (300 miles) away. Detonating several miles up, the blast laid waste hundreds of square miles of forest, with fallen seared trunks pointing away from the direction of impact. There was no crater. The whole thing, most likely an "ordinary" stone, disappeared in the explosions, leaving only dust behind.

It will surely happen again, as potentially hazardous asteroids cross the Earth's orbit all the time (not to mention comets, whose scary natures will be exposed in the next chapter). Hardly a year goes by without some kind of warning, or a sighting of a smallish one passing by (as in "there it goes . . ."), some even gliding between us and the Moon. The most recent close call came in 1982 from a daytime house-sized fireball meteor. Seen from the southern United States to Canada, it zipped through our atmosphere and then barreled back out into space.

The question is not so much about if they will hit, but what we can do about it. For starters, find out where the big ones are, hence the several dedicated searches for potentially hazardous asteroids to turn "there it goes" into "here it comes." Eventually (and we have time, since giant impacts are rare events), we can probably deflect any that take a bead on Earth, thus saving ourselves from global catastrophe. Blowing them up is the most obvious ploy (and perhaps least productive, since then you create *lots* of incoming bodies). More creative ideas involve using a small probe that acts as a gravitational tugboat, or painting one side white so that sunlight, absorbed on one side, reflected on the other, can do the same job. There's a whole industry waiting to be explored.

Pitch and Catch

Every planet in the Solar System is subject to impacts. The Moon bears the scars of a "heavy bombardment" that took place in the first half-billion years of the Solar System, when much of the left-over debris of formation was being swept up (as do the solid surfaces of Mercury, Mars, and those of most of the outer planetary satellites). With no air, no water, no erosion processes (except from heating and cooling cycles plus more impacts), the Moon has preserved the ancient record. Yet even now (like Earth), it gets the occasional hit. Estimating from the surroundings and absolute ages of rocks brought back by astronauts, the giant crater Copernicus, nearly 100 kilometers across, is less than a billion years old. Similarly sized Tycho may go back only 100 million years (well within the range of species-killing Earth impacts). Spreading out from Tycho is a sunburst of "rays" that cross the lunar disk and that were caused by ejecta digging lighter-colored rock from beneath the sun-darkened dusty surface.

If giant impacts are so powerful that the ejecta can go halfway around the Moon, it takes little imagination to realize that they can be launched off it altogether and toward Earth. The Moon, after all, is a relatively small body, and the velocity needed to escape gravity's grip (such that the body will not come back) is but

40 percent Earth's. Nearly 100 meteorites have been identified as coming from the Moon. They give themselves away by their chemical compositions, which are similar to those of the rocks returned by the Apollo missions. The dark spots seen on the lunar surface with the naked eye are gigantic, lava-filled impact basins that were formed more than 3 billion years ago. The Earth must have been subject to an incredible bombardment from their ejecta, but the record is so old it is long lost.

If the Moon, why not other planets? Mars is not all that big either, and thus also has a low escape speed. Moreover, it is much closer to the asteroid belt and subject to more of the potentially hazardous crowd. We find Martian meteorites as well, about three dozen known. Blasted off Mars's surface, they go into orbit around the Sun, where they are subject to cosmic ray bombardment, whose tracks can tell us how long the journey has been (millions of years) from the red planet to the blue-green one. We know they are from Mars because the chemical ratios in the trapped gases are similar to those known (from landers) to exist in the Martian atmosphere. One, "ALH 84001," found in 1984 in Antarctica, where the ice sheets trap vast numbers of meteorites, is an ancient volcanic rock about 6 inches across. When cut open, it seemed at first to contain "microfossils," and was touted as evidence of Martian life. Alas, the structures now seem to be of geological origin.

For all our expertise in space travel, we have yet to bring back pieces from a celestial body other than the Moon, and through the Stardust experiment, dust from a comet. Nature, instead, has done it for us. And if the Moon and Mars, why not Venus and Mercury too, or the satellites of Jupiter, even Saturn? Hard—if not impossible—to recognize, a piece of another planet may pass under your foot as you walk a mountain trail. All compliments of Jupiter, our real thunderbolt-tossing Zeus, and the inexorable force of gravity. And if these planets, why not Earth? We may have spent the past 4 billion years launching ourselves into space to be caught by nearby planets. Some visionaries have even suggested that we are spreading life out into the cosmos to be nurtured and grown elsewhere. A stretch of imagination, but who really knows?

We cannot, however, separate asteroids from comets. The effects of the two can be difficult to untangle, and there is clearly an overlap. Within the cometary realm, to which we now turn, not only does Jupiter get into the act, but so do the distant outer planets—and even the stars themselves.

Chapter 6

Crashing Comets

Fear. A terrible comet, trailing a long lustrous tail, works its way across the sky, bringing doom, death, destruction in its wake. Halley's comet appears and the Normans invade England (bad for Harold, not so bad for William). It comes back centuries later in our own time, and Mexico City is shattered by a great earthquake. And therein lies the

Tale of the Chainsaw.

My colleague Lew gave me My First Chainsaw (a tool as dangerous as any comet), a small electric. Upon bringing it into the office, we decided to test it out before I took it home. Lew plugged it in. I pulled the trigger and the lights went out. Goodness. In fact the power to the entire Astronomy Building went out. Heavens! We then learned that the whole north end of *campus* had gone dark. What power we wielded with a simple chainsaw. With a larger one we might conquer the World!

It was not to be. I no longer recall the reason for the sudden electrical collapse; it may have been the usual "squirrel in the transformer." Whatever it was, it was not we who had done the deed. The events happened at the same time, by coincidence, without cause and effect. "Post hoc, ergo propter hoc," Latin for "after this, because of this," reveals an insidious problem. Acceptance of the five words has led to terrible science. Or to the fear of comets as harbingers of disaster. There are so many bad things—wars, natural disasters—going on, and have been since remembered time, that whenever a bright comet comes by we can always relate it to one of them. Post hoc . . .

We now know the natures and origins of these "hairy stars" ("coma" Latin for "hair"). While they have long influenced humankind through the fear they have historically evoked (one storied regent apparently dying of a heart attack out of sheer terror, thus satisfying the prediction that he would die upon visitation by a great comet), they are both beautiful and mostly benign. Indeed, they *can* bring doom and destruction. But not by some mysterious nature. They instead have to hit us.

They also gave us life.

A Tale of Two Tails

A bright comet is one of the great naked-eye sights of the celestial vault, one that once seen is never forgotten. Comets are also the source of one of the great astronomical misconceptions (just behind the cartoonist's view of the telescope sticking out of the observatory dome). Bad movies and more cartoons have comets streaking furiously across the sky from one horizon to the other, confusing them with meteors, which do indeed so streak. Instead, comets are interplanetary bodies that orbit the Sun and that, like planets, move leisurely against the stellar background, slightly displacing themselves from one night to the next. The difference is that comets move on highly elliptical paths that can take them both far from the Sun then right up close to it, whereas planetary orbits are more or less circular. Meteors, on the other hand, are nearby atmospheric phenomena. Giving some excuse for the misconception, most meteors—tiny rocks that hit us and usually burn up in the air—are the comets' spawn.

Before we can make the connection, however, we need look more closely at the comets themselves. Dozens of new (and returning) comets are found every year, the vast majority being small ones visible only through telescopes. Big naked-eye comets appear at random every decade or two. Until the recent advent of large

survey telescopes, most comets were discovered by dedicated amateurs who set out to find them. Scanning the skies near sunset or
sunrise with a wide-angle telescope or even binoculars, the finder's
reward is to have his or her name attached to the cosmic journeyer.
Comets also carry more formal alphanumerical designations that
give the year, even month, of discovery.

Comets are "head and tail" affairs. At the center of the bright head
is a tiny nucleus that contains nearly all the mass and that creates
a surrounding fuzzy nearly opaque "coma," from which streams
not one, but *two* glowing tails. The nucleus is so small, only tens of
kilometers across at best, that it appears (unless overwhelmed by
the glowing coma) as a bright, unresolved point of light. The coma,
on the other hand can extend over several degrees.

The tails do not behave like a thrown rock with an attached
paper streamer. One of them, long and thin with a bluish color,
always points directly away from the Sun no matter what the comet's motion. It can as well lead the comet in orbit as fall behind it.
The other is broad, white, and tends to curve. While often pointing
away from the Sun, it can just as easily seem to point toward it.
Which tail dominates depends on the particular comet, and sometimes only one of them is clearly visible. At times the tails are short
and stubby, while in other cases they may stretch across the entire
sky. Perspective adds to the mix. The closer the comet is to the Sun,
the brighter it becomes and the longer its tails, showing who really
runs the whole affair.

Comets' natures are resolved by spectroscopy, in which the cometary light is spread into its spectrum, or component colors. From
the coma we see colored emissions identified as radiating from
simple molecules such as carbon monoxide (CO), molecular carbon (C_2, C_3), cyanogen (CN), water (H_2O), and others. Underlying
the emissions is the reflected spectrum of sunlight.

The tails separate the two kinds of light, the blue tail showing
only the emissions, the white one only the solar reflection. The
emissions are a giveaway that the coma, and especially the blue
tail, are gaseous. Since some of the molecules are ionized (electrons
stripped away), the gas tail is equally well known as the "ion tail."
Reflected sunlight, on the other hand, can come only from solids.

Figure 6.1. Hale-Bopp, one of the great comets of the twentieth century, soared across the sky in 1997. Rarely do comets so beautifully separate such prominent straight ion tails (the lower, fainter of the two) from their broad, more curved, dust tails. Gas tails can be up to an Astronomical Unit long. Chaotic ejection from the nucleus is seen in the fine, "shredded," structure within the dust tail. Courtesy of A. Dimai and D. Ghirardo (Col Drusciè Observatory), AAC.

Since you can see stars right through the white tail, it must be made of fine dust grains (which are also embedded in the coma), and is hence called the "dust tail." Sometimes the dust tail splits into parts, giving the comet a spectacular set of multiple tails.

A Scary Tail Tale

The most famed of comets is named after Edmund Halley (1656–1742). A friend of Newton's, Halley was the first to calculate cometary orbits. He discovered that comets appearing in 1531, 1607, and 1682 had identical paths and were therefore likely to be the same one returning every 76 years. The highly tilted orbit takes the comet from as close as 0.59 Astronomical Units to the Sun to as far as 35 AU, past the orbit of Neptune. He then predicted the return in the winter of 1757, which came 15 years after his death. Appearing just a year behind schedule, Halley's Comet became his eternal memorial. It has been traced in written records as far back as 240 BC.

Halley's then showed up in 1835 to mark Mark Twain's birth, and again in 1910 to make note of his death. That passage was one of the best ever, the comet coming close to the Earth while at the same time it was closest to the Sun and most active. It created a sensation, caused not in small part by the sky's darkness, unpolluted by city lights. Cartoonists had great fun showing the comet smashing into Earth, its "gravity" (or whatever mysterious forces came from the imagination) tearing down buildings, the evil apparition even carrying off small babies. On the plus side, it inspired a "Comet Rag" for piano composed by one C. Mahoney and a "Comet March" by C. E. Duffield.

The scary part of this particular tail tale involves more cometary doom and destruction. On the night of May 18, 1910, Earth was predicted to pass through the comet's long tail. That would not have been so bad, except that shortly before Halley's return, the deadly gas cyanogen had been discovered in other comets' tails. In spite of astronomers' assurances to the contrary (comet tails being remarkably thin), some feared sickness or even poisoning. People stuffed wet towels around their doors, hoping to protect themselves against the "invisible killer," while—given that it might be their last night on Earth—city sophisticates held parties. Nothing happened. The sky did not even glow any brighter.

Things were a lot quieter in 1985–86, when Halley's came back again in one of the worst appearances of the last two millennia, the comet near perihelion while on the other side of the Sun from us and about as far away as possible. Given the hyper-publicity of the return, many were those who saw an evening contrail from a jet aircraft and were fooled into thinking they had seen the comet's tail. Nevertheless, glorious it was from the Earth's deep southern hemisphere. Be sure to watch for it in 2061–62 (and have an aphelion party in 2024, when it will be farthest from the Sun).

These beautiful and mysterious phenomena have a surprisingly homely explanation as huge cosmic snowballs.

Structure

Fred Whipple of Harvard looked at the data and created the standard model. Greatly simplified, comets are watery (icy) "dirty snowballs" on long looping orbits that take them from the deep freezer of the outer Solar System into the comparative furnace of the inner. You can make a simple one yourself by going to Chicago or New York in February and scooping snow from a gutter that has absorbed the dirt and chemicals of a thousand buses and taxicabs.

Much of a comet's mass is made of ordinary water ice, which is mixed in with a variety of other stuff that includes carbon and nitrogen compounds and complex organics. The ices provide a matrix that embraces rock and fine grains of dust. In order to have such a high water content, comets had to be formed as planetesimals far from the heating Sun and only later brought inward by the gravitational action of the giant planets or even by that of nearby passing stars.

As a comet's nucleus nears the Sun, solar radiation heats the ices and evaporates them to a gas that is broken down to simpler molecules and then ionized by sunlight and the solar wind to create the coma. The solar wind, which continuously pulls the Sun's magnetic field outward, slams into the developing ionized coma. Wrapping the solar field around the comet like a windsock, the wind and the solar field push the cometary gases downwind to create the arrow-straight gas tail. Indeed, the very existence of comet tails was one of the first indicators of the solar wind.

The Sun's rotation spins the solar magnetic field into a huge expanding spiral with some of the field lines pointing outward, some inward to preserve magnetic balance. When the comet passes from an outward-pointing sector to an inward-pointing one (or vice versa), the gas tail snaps off and drifts away, while a new one forms. These "disconnection events" are sometimes even visible to the naked eye.

As the ices melt, dust and small bits of rock are released to join the gases of the coma. Not ionized and therefore not affected by the solar wind, the dust grains, most a thousandth of a millimeter or so in diameter and akin to smoke, are dragged away from the comet by the pressure of impacting sunlight. The dust, having more

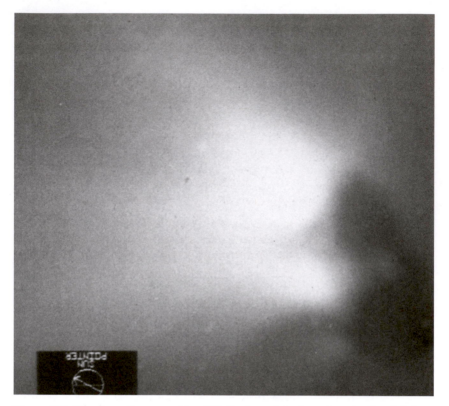

Figure 6.2. The heated, sunny side of Halley's coal-dark nucleus (at far right) sprays fountains of dusty gas into space. The 16 × 8 kilometer (10 × 5 mile) "peanut" will quite likely break in half and eventually waste away. European Space Agency, Giotto Camera Team.

mass than an atom or molecule of gas, also displays much more inertia. Depending on the comet's path and direction at the moment, the dusty spray therefore curves in a direction different from the gas tail to form the white, reflective dust tail. It can even seem to point toward the Sun. Meanwhile, the larger chips of mostly invisible rock spread outward.

Simplicity is for the human mind. In reality, nature is always far more complicated, if not enigmatic. At Halley's last return, it was intercepted by a variety of spacecraft, one of which managed to get a fairly detailed close-up image. The peanut-shaped mass, only 8 × 16 kilometers across, is extremely dark, reflecting about as much light as black velvet. Over long periods of time, cosmic rays (energetic particles from interstellar space accelerated by exploding stars) have acted on the surface to create dark, complex molecular tars of an unknown nature.

Supporting a long series of ground-based observations, the spacecraft showed that the gas does not evaporate uniformly from the surface, but emerges from deep inside in powerful fountains that spiral around as the nucleus rotates and that entrain the dust. Dust falling back to the nucleus can then compact into larger pieces of fluffy rock that get wrenched off by the more energetic jets. The jets are even strong enough to act as small rockets that alter the comet's orbit and make its future path rather unpredictable. Gravitational actions of other planets constantly change the orbit as well. As a result, a comet often comes in toward the Sun on one orbital path and goes out on quite another.

Getting Back (Twice)

Comets do—and will—reach out to touch Earth via violent collision. To understand comets and the dangers they pose, it seemed only fair to smack one of *them* instead to see what is inside. In the summer of 2005 the *Deep Impact* spacecraft violently visited Comet Tempel 1, which orbits the Sun every 5.5 years. Images revealed an irregular rocky body 5 × 7 kilometers across covered with both craters and strange, totally smooth "plains." On July 3, the spacecraft fired a refrigerator-sized, third-of-a-ton "smart bullet" (with its own imaging and guidance capability) toward the comet. A day later, the scientific artillery collided at 5 kilometers per second (11,000 mph). With a flash visible on Earth, it ripped out a stadium-sized crater, in the process kicking away a huge amount of icy dust for the passing mother ship (watching from a safe distance of 500 kilometers) to observe with a variety of instruments.

Examination of the debris revealed a lightweight body with a density a mere six-tenths that of water, just 20 percent that of Earth rock (similar to that estimated for Halley's). The amount and velocity of the ejecta suggested that the comet was not held together by the strength of its rock, but by its weak gravity. It's (again) a pile of rubble. Consistent with Whipple's early ideas, it's also layered, with a lot of ice that was directly observed even as surface patches. Mixed in were carbon dioxide, hydrogen cyanide, methyl cyanide, various organics, and actual rock (carbonates and iron-rich silicates). Most surprising were "clays," which required the presence of liquid water at some time in the past. Equally tantalizing, we have no idea how typical this comet might be, leaving an amazingly rich field for future exploration.

The *Stardust* spacecraft, launched in early 1999, gave a whole new meaning to "getting back." Taking four years to get to Comet Wild 2 ("Vild," which orbits roughly between Mars and Jupiter), the craft carried a panel covered with a sticky "aerogel" designed to pick up the comet's flaked-off dust particles. In January 2006, the spaceship returned to Earth with its load, dropping the package of precious particles to our planet by parachute. While most are microscopic, a few (hammered firmly into the aerogel) can be seen with the naked eye. Though comets are supposed to have been created in distant space beyond Jupiter, where ice can be trapped by growing planetesimals, they show evidence of rock types formed under high temperature conditions close to the Sun. Nobody knows why. As in most areas of astronomy, the more we know, the less we seem to understand.

Breaking Up (Is Not That Hard to Do)

If full-blown comets (those that come near the Sun with coma and tails) are evaporating, they must be temporary creations of nature. Losing perhaps a percent of their mass at each perihelion passage, they are all evaporating away, their lifetimes dependent on their individual structures, their orbital periods, and how close they come to the Sun.

Comets' fragile natures are quickly revealed not only by spacecraft, but also by their breakups. Comet West of 1976 went

Figure 6.3. As short-period Comet Schwassman-Wachmann-3 toured the Sun in 2006, it broke into pieces (one of which is seen here) that further shattered into even more pieces. It may come back as many small comets, or perhaps never be seen again. Hubble Space Telescope: NASA. ESA, H. Weaver (JHU/APL), M. Mutchler, and Z. Levay (STScI).

around the backside of the Sun and cracked into four or five separate pieces, all of which will someday return as separate, smaller comets. In May 2006, Schwassmann-Wachmann-3 crumbled into dozens of tiny chunks. The champion breakup act belongs to the "Kreutz group" of close-in "sungrazers." They consist of a huge number of pieces of some dim-past parent body that have been coming back for over a century; some of them plow right into the Sun itself. Collision. Something to ponder.

Rather than disintegrating, however, some comets seem simply to dry up. Losing their tails, they masquerade as asteroids with highly elliptical orbits. To confuse the issue further, several "asteroids" have been seen to develop unexpected tails, showing that they either are "wet" or are real comets. The forms may overlap. Whether still whole, falling apart, or dried and defunct, comets trail clouds of dust and rocks. Like rainclouds coming in over the horizon, they are responsible for a forecast of

"Passing Showers."

Cometary disintegration can be seen any clear night. While aster-oids produce an occasional meteor, and perhaps most big fireballs, the vast majority of meteors is caused by comet fluff. Under a fully dark sky in the early morning, when you are on the forward side of orbiting Earth, you will on the average see at least three an hour—the base number—zipping across the sky. Evidence for cometary origin lies in occasional meteor showers.

On the morning of August 12 or 13, the average number of me-teors goes up to close to 100 or more an hour, two a minute. If you chart the paths, they all trace back to an origin in the constellation Perseus, so as a group are called the "Perseids," the most famed shower of them all. In early January we see the Quadrantids (from their origin in the defunct constellation Quadrans, the Quadrant), in mid-December the Geminids, both of which can also produce 100 an hour. We are *always* in some kind of shower, though it may be so sparse as to be recognizable only by experts. Under a moon-less dark country sky, the silent fireworks display is a celestial art-work that stirs the soul, allowing us to ponder our relations with the cosmos.

Many of the showers appear when we pass by, or even closely approach, the orbit of a known comet. The relation is so common that we may safely assume that *all* showers are related to comets. That there may be no related comet just means that the progenitor has died away or has such a long period that its last return was lost to history. The Perseids are associated with comet Swift-Tuttle, which orbits the Sun every 130 years and last passed us in 1992. The Geminids come from what was once *thought* to be an asteroid, but has since been identified as a dried-out comet. We can thank Halley's for *two* showers, as we approach its orbit twice during the year, first around May 5 for the "Eta Aquarids" (named after a star in Aquarius), then around October 21 for the "Orionids" that come out of Orion (each producing 20 or so per hour). Most showers take a week or more to build up to their peaks, which can be quite sharp. Then they gradually fade away, only to re-appear the following year.

The showers are caused by the larger solid particles that are ejected from the comet's nucleus (the dust grains too fine to produce visible events). Ranging from a few millimeters to perhaps a meter across, the rocky clumps that hit us are big enough to set up visible atmospheric shock waves that mark the particles' meteoric paths. The meteors are subject to a strong perspective effect. Pieces of cometary junk orbit the Sun on more or less parallel paths. When the Earth runs into the cloud of debris, we see the meteors appearing to come from a divergent point much as parallel railroad tracks seem to come from a point over the horizon. The direction of the divergent point depends on the relative motions of Earth and comet. As time passes, the particles slowly migrate from their original cometary orbits. Their origins lost, they make the sporadic meteors seen on any clear night.

The "Mother of Them All" is the Leonid shower (from the constellation Leo) that usually peaks the morning of November 17. It's related to relatively faint Comet Tempel-Tuttle, which orbits the Sun every 33 years. The shower is usually drab. But every 33 years, as the comet travels past us followed by its stringy cloud of debris, we may see a rare "meteor storm" in which up to 100,000 an hour can light Earth's upper atmosphere. Unfortunately, the

Figure 6.4. At left, the 1976 Leonid shower sent dozens of meteors across the line of sight, their paths clearly diverging from a point far up and to the right. At right, we look right down the pipe at the divergent point itself, seen against the constellation Leo (its bright star Regulus toward lower right). NOAO/AURA/NSF/WIYN.

blob of stuff is continuously perturbed by Jupiter, so the storm may or may not appear. Drawings made of the spectacular 1833 event (which to those who saw it seemed to presage the end of the world) show the sky filled with meteors, a celestial snowstorm. The 1966 storm was spectacular, while the one at the end of the twentieth century was good, but not grand. No matter what the shower, Perseid, Leonid, or obscure, there is no increase in the rate of meteorite falls. None of the delicate bits ever hits the ground. Watch and enjoy. Collision insurance not necessary.

Twilight of the Dust

As the Sun drops below the horizon and the ground darkens, sunlight continues to play off the atmosphere, giving us a period of ever-deepening twilight. Eventually the Sun glides so far down (by technical definition, 18 degrees below the horizon) that even the upper air cannot "see" it any more and the sky becomes fully dark.

But not necessarily. If there is no Moon, no surrounding lights, and the air is brilliantly clear, you may see a cone of light standing up from the western horizon, a seeming memory of departed twilight. Not of Earth, it is caused by the scattering of sunlight from the countless dust particles that lie in the plane of the Solar System. Centered on the ecliptic (the apparent path of the Sun), and thus on the constellations of the Zodiac, the "zodiacal light" fades away, only to reappear in the morning sky as the "false dawn" that precedes the beginning of real twilight.

Exactly opposite the Sun, high at midnight, is a faint glowing "gegenschein" (German for "counterglow"), caused by extra-efficient backward scattering of sunlight by these same particles. During years of solar quietude, when the solar wind blows softly and the airglow is much reduced, you can see the zodiacal light arching across the entire celestial vault like a second, albeit much fainter, Milky Way.

You are seeing the debris of disintegrated comets and smashed asteroids that slowly sweeps inward under the radiative action of sunlight. Which kind of object contributes more depends on the cloud from the latest major asteroid collision to reach us or the latest great comet or comet shower to pass us. The zodiacal light thus fuses the two forms of ancient planetesimal together in a mix of cosmic demise. Swept up by Earth, some of it never gets to the Sun, but instead filters down, falls to the ground in cleansing rain, the rain from the heavens themselves blending into the watery fall, asteroids and comets becoming part of our own planet on which we walk.

Pathways

Invading comets are quickly destroyed. Yet four and a half billion years after the birth of our Solar System, comets seem everywhere. There must be a huge supply of them. To find their origins look to their orbits. Comets also display a great variety of pathways. The short-period record is 3.3 years for Encke's comet, its highly eccentric path taking it roughly between the orbits of Mercury and Jupiter. Encke's obviously has little time left before it either disintegrates or looks like an asteroid. The long-period limit is unknown. Comet West will not return (don't leave the light on) for half a million years, which would take it out to a distance of over 12,000 Astronomical Units, 300 times farther than Pluto.

Some periods are much longer. Stop reading for the moment. Get some exercise by throwing a ball in the air. Under the action of gravity (you are still reading?), it slows down, stops, and then comes back. However, if you throw it up at greater than the "escape velocity" (11.2 kilometers, 7.1 miles, per second), though always slowing down, it's now going too fast to come to a halt, and it leaves forever. The track taken by a body moving relative to another with a speed greater than the escape velocity is no longer elliptical, but has the shape of an open-ended *hyperbola*: Newton's variation on Kepler's last law in action. The curve that separates the hyperbolae from the ellipses is the *parabola*, in which a thrown body is exactly at the escape speed. It will come to a halt, but only after an infinite period of time.

At the extreme, comet orbits take on shapes close to that of a parabola. But there are none at all with the clear hyperbolic orbits that would imply that they are coming in on one-way orbits from interstellar space, from other stars. The near-parabolic orbits must really be ellipses so hugely extended that, given the accuracy of the data, they just appear as parabolae. All known comets therefore belong to the Sun.

There is not, however, a continuum between longer and shorter period comets. The two are different. A vague dividing line appears at a period of a couple hundred years. Shorter-period comets tend to stick to the Zodiac and to orbit the Sun in the same counter-clockwise direction in which the planets move, while longer-period ones can be found anywhere, with just as many going the "wrong way" as the planetary "right way." Moreover, the really bright ones, those that live in memory (Halley's an exception), are of the long-period sort. The two groups, while sharing many physical characteristics, must have different wellsprings.

Kuiper

Gerard Kuiper (1905-1973), a Dutch-born American astronomer, helped figure it out. Following an earlier suggestion by K. E. Edge-worth (Ireland, 1880-1972), Kuiper conjectured that to account for the distribution and orbital direction of short-period comets, they must be coming from a extended disk of icy bodies that lies near and beyond the orbit of Neptune, one that shares the general plane of the Solar System. Given the number that invade the inner Solar System and the estimated destruction rate, there must be hundreds of millions, perhaps billions of them.

Lurking in the frozen depths of space far from the Sun, they stayed hidden until a growing telescopic technology found the first one, a historic sighting of drably named 1992 QB1 (year of discovery followed by the order within the year), which averages 43 AU from the Sun and was estimated from its brightness to be only around 240 kilometers (150 miles) across. By the end of 1995, two dozen Kuiper Belt Objects (KBOs) had been found, all between 40 and 46 AU out. The number now exceeds 1000. All of a sudden, Pluto began to make sense. A tiny body (2300 kilometers, 1400

Oort

Shortly before Kuiper made his suggestion, the Dutch astronomer Jan Oort (1900–1992) offered a somewhat similar explanation for long-period comets. For there to be as many as we see, with such great orbital periods that are so evenly distributed about the sky, the comets must originate from a huge cloud—the "Oort Comet Cloud"—that extends to at least 50,000 AU from the Sun, and perhaps halfway to the nearest star (which lies some 300,000 AU off). Given the rates of visitation and destruction, there may be a trillion cometary bodies in the Oort Cloud, far more than in the Kuiper Belt. Yet they are so small that their combined mass may be only a few tens of times that of Earth.

Oort comets are so distant that, unlike those of the closer KBOs, we have yet to see and confirm any "in place" rather than as they enter the inner Solar System. Halley's, which unlike Kuiper Belt comets, orbits backwards and way out of the plane of the planetary system, is probably an Oort Cloud comet that got captured into a tighter path by planetary gravity. Sedna may have belonged to the Cloud as well and might thus represent its innermost extension. For all we have learned, however, some two centuries after Herschel expanded the Solar System with the discovery of Uranus, we still do not know the System's full inventory.

There and Back

Ignoring overlap between the two kinds of planetesimal, the dry leftovers near the Sun became the asteroids, while the "wet ones" farther out turned into comets. Just after their formation, theory suggests that Uranus and Neptune tossed huge numbers of leftover cometary bodies outward to populate a growing Oort Cloud, while Jupiter and Saturn not only added to the mix, but probably hurled most of them out of the Planetary System altogether.

Giant planets have the gravitational power to move comets and asteroids, and can even shift the Earth, changing its orbit. But gravity is always mutual, and given enough time, small bodies can also affect the big ones, the degree depending on the big body's mass. An interaction with Jupiter that tosses a planetesimal outward at high speed causes the planet to move slightly inward from its birthplace. Saturn moved outward some, as did Uranus, while Neptune seems to have spiraled outward by some 10 Astronomical Units! As it slowly tractored along, it also shoveled remaining small bodies ahead of it, creating the current Kuiper Belt and capturing some of the KBOs into their 3:2 Plutino orbits. Pluto included.

Next, we have to get the comets back into the inner Solar System on solitary visits. We can see the Kuipers coming. Between Jupiter and Neptune is a thinly populated "asteroid belt" of about 100 known objects in eccentric inclined orbits that have been sent there through mutual collisions and planetary perturbations. Some of them cross the planetary paths themselves. While the first one found averages 13.7 Astronomical Units out, it comes inside the orbit of Saturn. Discovered in 1977, it was named Chiron after the classic centaur who mentored Jason prior to his quest of the golden fleece. In its honor, they are collectively labeled "centaurs." When Chiron moves closer to the Sun, it develops a gaseous coma: it seems to be a giant comet 100 or so kilometers in diameter! Huge numbers of smaller, yet undetected, ones undoubtedly exist.

Centaur orbits are unstable. As the small bodies approach and then cross the orbits of the giant planets, gravity moves them ever inward. Think of it as a scene from a bad horror movie with zombies lurching inexorably toward the unsuspecting villagers: us. Smaller centaurs eventually appear as simple short-period comets. Large ones are clearly rare, but must be spectacular when they first approach the Sun.

A tiny fraction of Oort Cloud comets suffer similar fates. Most will forever mind their own business far from the Sun, at distances of hundreds, thousands, even tens of thousands of Astronomical Units. But a few get back here, amazingly enough through the actions of the stars. Stars do not collide in our part of the Galaxy. The nearest is 30 million times the solar diameter, making such collisions a statistical near-impossibility. But stars can certainly come close to, if not enter, the gigantic Oort Cloud. So can massive blobs of star-forming interstellar gas and dust. Such encounters can alter the distant cometary orbits, whence a few will be dropped into the inner Solar System to create the next Great Comet, another Hale-Bopp (of 1997) perhaps. Even tides raised in the Oort Cloud by the Galaxy itself can have the same effect, bringing us ever closer to the distant cosmos.

Collision

Space is big, the Earth no more than—in the words of Carl Sagan—"a pale blue dot." Time, however, is on the side of the comets. On occasion, and no one knows the rate, one will—like an accidental asteroid—collide with us and visit devastation on Earth and its various forms of life. Can it really happen? While the world watched with fascination, during the third week of July 1994, Jupiter took a direct hit from Comet Shoemaker-Levy 9. Sometime within the past couple hundred years, this small comet, probably flung in from the Kuiper Belt, passed too close to Jupiter and its satellites. Stolen from the Sun, it became trapped into a wide Jovian orbit. Not entirely giving up control, the Sun then pulled it into an ever-skinnier path. In July 1992 (so we surmise from its orbit), the comet came so close to the giant planet that Jovian-raised tides broke it apart into pieces each less than a few hundred meters across. By the time Carolyn and Eugene Shoemaker and David Levy found it in March 1994, it had fragmented into 22 segments, each with dusty tails following the other. As the orbit narrowed yet further, Jupiter just got in the way. Over a six-day period beginning on July 16, one after the other plowed into the planet's deep atmosphere.

Even though the actual collisions took place on Jupiter's night-time side facing away from Earth (perversely not letting us see them), the results were still spectacular. Ramming into the planet at 60 kilometers per second (135,000 miles per hour!), the bigger pieces blasted brilliant infrared fireballs that rose to over 3000 kilometers above Jupiter's cloud-covered surface (half the width of the continental United States) loaded with stuff from Jupiter's

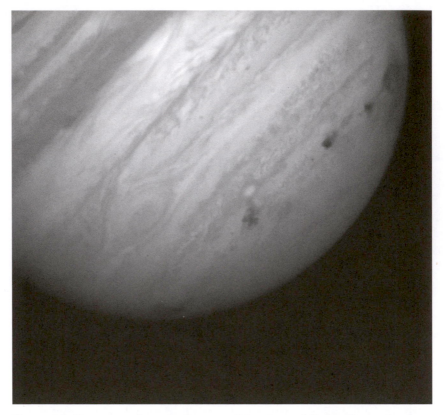

Figure 6.6. Dusty scars from the impacts of the pieces of Comet Shoemaker-Levy 9 litter Jupiter's southern hemisphere. Jupiter's ammonia clouds are colored by various hydrocarbons, while the Great Red Spot, a rotating "anti-hurricane" twice the size of Earth, lurks to the left. Hubble Space Telescope Comet Team and NASA.

various inner cloud decks, including sulfur compounds and water. As the Jovian skies cleared, we saw the planet marked with dark dusty scars, some bigger than Earth, that lasted for weeks until fierce winds finally stretched them into oblivion.

The total energy release was the equivalent of 25,000 megatons of TNT, 250 times that of the largest hydrogen bomb ever conceived and the rough equivalent of the world's nuclear arsenal. All from a parent body that is estimated to have been no more than 1.5 kilometers—a mile—across. Searches for historic observation of dark spots on Jupiter and Saturn suggest that it has happened before. If it could happen out there, it could happen here.

Strikes and More Showers

There are too many comets coming close to the Sun for us not to have been a successful target. The great Tunguska impact was long thought to have been a comet strike, though the evidence from embedded particles now leads strongly toward an asteroid. That the Chicxulub (Yucatan) impact may have been caused by a comet cannot be ruled out.

While both can create havoc, there is a distinct difference in the encounters between asteroids and comets. Given their closeness, we can at least see asteroids approaching us long before they are dangers. If we can find them, we can predict their orbits decades into the future, enough to give ample warning. The only problem is initial discovery, and that is being solved by dedicated searches. Kuiper Belt comets are vaguely similar. But not those of the Oort Cloud. Dropping suddenly into the inner Solar System at high speed from immense distances, they give us little notice, only a couple years perhaps. We will have to have "quick response" capability to deal with them. And at this point, even "slow response" is lacking.

Like their meteors, comets can also come in showers, perhaps even in storms. The gravity of a star or interstellar cloud passing within the confines of the Oort Cloud, or even one that comes close, will most likely not just send one comet pounding down its new orbit toward Earth, but whole bunches of them, greatly increasing the odds of a strike.

But the phenomenon goes far beyond the invasion of the Oort Cloud by single stars. We live toward the outskirts of a large spiral galaxy some 100,000 light-years across. The light-year, the distance light travels in a year at 300,000 kilometers per second, is 63,000 Astronomical Units, 10 trillion kilometers, long. The vast bulk of the Galaxy's 200 billion or more stars, which in our neck of the woods average a few light-years apart, lie in a thin disk that includes our Sun and is seen around us as the Milky Way. The disk, which contains a graceful set of spiral arms that sweep out from the center, is rotating such as to wind the arms up. But since the arms are really waves of increased star density (number of stars per cubic light-year), new arms form as old ones disappear.

Partaking of the general rotation, the Sun orbits the Galactic center with a speed of 220 kilometers per second, and at our distance from the center of 25,000 light-years takes about 200 million years to make a circuit. The orbit is slightly inclined to take us a bit above the Galaxy's dense center line to slightly below and then back again. As the Sun orbits, it continually passes through the main disk and also in and out of spiral arms. An especially high star density will have a more powerful effect on the Oort Cloud's comets, as will tides in the Cloud that are raised by the whole stellar collection. In times of such passages, we might expect a greatly increased comet fall. As on Earth itself, there is always stormy weather ahead.

A comet shower may have been responsible for the late Eocene mass extinction of 35 million years ago. The extinction event coincides with the dates of two major impacts. Creating a crater 90 kilometers across, the Chesapeake Bay strike hurled tektites from northern Virginia to Georgia. With a diameter of 100 kilometers, Russia's Popigai crater is contemporaneous and slightly larger. Iridium levels of the time are also high. Were the pair a result of a double asteroid or comet shower?

To decide, look at what the dust dragged in. Helium, the second lightest chemical element, comes in two forms. Atoms of normal helium-4, the stuff that floats helium balloons, have two protons and two neutrons. The other kind, helium-3, has but one neutron per atom. Because of its lightness and volatility, virtually no helium was incorporated into the primitive Earth. Nearly all we have was produced by the radioactive decay of uranium, which spawns only helium-4. Cosmically, however, including in the solar wind, about one helium atom in ten thousand is the lighter version.

Interplanetary dust from comets and asteroids picks up both kinds of helium from the solar wind. Whatever little helium-3 is owned by Earth has been incorporated by accretion of these tiny dust grains. A rock layer relatively rich in He-3 therefore reveals a cosmic dust shower hitting us. From a million years before the extinction and the two cratering events to more than a million years afterward, rock layers are filled with helium-3, revealing an associated dust shower.

Asteroid collisions produce families of smaller bodies, some of which may be sent inward to impact Earth. They also produce dust clouds that take much longer to get here. Dust associated with a comet shower, however, would arrive about at the same time as the comets. Which is just what we find.

But comets bring far more than trace elements. They also bring

Water.

All day I face the barren waste without the taste of water,
Cool water.

(Bob Nolan)

So sang the old Sons of the Pioneers. Desperate thirst. We seem to be here because cometary and asteroidal impacts left the Earth open for the development of new forms of life. Though far from proven, the age of dinosaurs was perhaps begun by the monstrous Antarctic impact, then ended (far more certainly) by Chicxulub, which in turn left the world to mammals, and ultimately to ourselves. But without that great solvent, water, no life would have been possible in the first place.

Water is volatile. Near the hot Sun, none of it should have been incorporated into early Earth. It could freeze onto a planetesimal surface only beyond the snowline in the outer asteroid belt near Jupiter, leaving the inner planets incredibly dry. They still are. Though Earth, covered with oceans, seems to be terribly wet, that is but an illusion, as the seas average a mere 3 or so kilometers deep on a planet 13,000 kilometers (8000 miles) across. Yet there is enough to support abundant life.

One explanation jumps forward. Once our primitive dry Earth cooled from the heat caused by the rapid-fire impacts that built it, the gift of water was given to us by the accretion of hoards of leftover comets that hit us, each sent inward by the outer planets, each incorporating a bit of water that could then stick around. Other volatiles arrived as well. The process goes on today, albeit at a much lower level. We—as humans—are not just products of Earth, but of the planetary system as well. Even of the whole cosmos.

Spacefarers

For years, centuries, we have dreamt of going to the stars. While in our own times we have sent four spacecraft into the cosmos (the two *Pioneers* and the two *Voyagers*, which will never return), nature has long done it for us. Like planets tossing rocks back and forth, so—one would think—must the stars.

If the giant planets indeed filled the Oort Cloud with leftover icy planetesimals, they must also—as theory suggests—have launched far more at speeds sufficient to have escaped the Sun altogether. Moreover, Oort Cloud comets are so loosely bound to the Sun that any disturbances that send us comet showers must also eject comets that will forever escape. We have been leaking comets into the Galaxy for nearly 5 billion years. In all that time, have some of them encountered other stars? They would enter other planetary systems on high-speed hyperbolic orbits and be instantly recognizable by local astronomers as visitors from afar. Perhaps one has even hit a planet in a distant stellar family.

It works both ways. Since our Sun and planetary system hardly seem unique, other stars must eject similar numbers of their own comets. If so, we might expect to see comets invading our system on similar fast orbits, not as fictional UFOs but as *real* "space aliens." We have yet to find any, though the number may be so small and we have looked for such a short time that their apparent lack is unsurprising. If we ever do find any, it will cause a sensation.

Comets thus provide our first real encounter with what is out there far beyond the Sun. But not the last, as using comets as our springboard, we now launch ourselves into deep space on a journey that will take us to the farthest reaches of the Universe, all of which is of a piece with our quiet and beautiful Earth.

Chapter 7

Atomic Rain

Faster than a speeding bullet; much more powerful (per gram) than a locomotive; able to leap broad galaxies in a single bound! Look, up in the sky. It's *cosmic rays*. And like neutrinos drifting down a dusty hall (with apologies to John Updike), you are again completely unaware of them. Or so you might think.

Radiation

A word that, for all our modernity, can still spark irrational fears that replace the terror once evoked by comets. While some kinds of radiation are indeed properly fearsome, most of it is not only a benign background, but necessary to life. The most familiar is the electromagnetic radiation of which ordinary light is a part, the visual radiation seen as violet through red. It stretches out along the electromagnetic spectrum to longer and less energetic waves as infrared and usually quite harmless radio. With some exceptions, as microwave in ovens, only the energetic waves shorter than violet light can do any real harm. Fortunately, most of the burning ultraviolet and X-ray radiation from the Sun are blocked by the Earth's atmosphere: else we would not be here.

"Radiation," however, has a double meaning. Scientific naming often precedes understanding (while at the same time *leading* to understanding). When early scientists began to study what we now call radioactive materials—radium, uranium, and the like—they found three kinds of natural, dangerous emissions that they called after the first three letters of the Greek alphabet: alpha, beta, and gamma. Continued investigation showed that gamma rays were the shortest-wave tail of ordinary light, while the other kinds were made of escaping particles: "beta rays" are energetic electrons, "alpha rays" high-speed helium nuclei. In deference to the past, helium nuclei are still commonly referred to as "alpha particles." Uranium, for example, turns itself into lead through a series of beta and alpha particle ejections. The process creates radium and radon (of household air pollution fame), and is vital for dating rocks.

At one point, "cosmic rays" were thought to be ultra-energetic electromagnetic radiation off the high end of gamma rays, and are still erroneously presented as such in a variety of badly written science books. Instead, they are a form of particle radiation. Not of Earth, they are produced by exploding stars and other high-energy phenomena, and involve not just our whole Galaxy, but galaxies beyond.

Without them, life and laundry would not be the same. Nor would our Sun and its planets. Indeed, none of these may ever have existed.

Laundry?

Stumbling in the Dark

Or, "How Science Often Works." As generations of bored students have reluctantly learned, scientists use a "scientific method" in which one first gathers data, then develops a theoretical model to explain them. The model in turn predicts new experiments or observations that can test it, which leads to a better theory, etc. It works well, and can be fun to practice, especially while disproving somebody else's theory. As much excitement, though, comes from pure accident. While running experiments to test your pet theory "X," you stumble blindly upon "Y," which you never knew was there, which in turn leads to a whole new direction of exploration and a Nobel Prize. An accident, yes, but by a scientist who knows what to make of it. A prime example of such scientific serendipity is chapter 4's "nutation," a set of small wobbles in the Earth's rotation axis. Nutation was found while looking for stellar parallaxes, the minute shifts in the positions of stars caused by the orbiting Earth, from which stellar distances are derived.

Cosmic rays, high-speed atomic nuclei that pervade all of outer space, provide an energetic example that really did lead to Nobel Prizes. Roaring through our Galaxy, interactions with quieter interstellar atoms have completely stripped them of their surrounding electrons, leaving them as pure, positively charged ions. The great majority are simple protons, or hydrogen nuclei: no surprise since hydrogen makes 90 percent of all the Universe's matter. "High-speed" is hardly an adequate description. Most cosmic rays zip through interstellar—even intergalactic—space at velocities greater than 99 percent that of light, giving them immense energies.

As they pound toward us, they first encounter our Solar System at a distance of about 100 Astronomical Units, where as charged particles they must climb over the barrier built where the solar wind and magnetic field ram into the gases of interstellar space to carve out our local bubble, the "heliosphere." The more energetic ones that get through then must further navigate the Earth's magnetic field, which cuts off even more of the lower-energy end. Finally, they all meet their match when they smack into Earth's atmosphere, which stops them cold.

But not without after-effects. The energies of attack are awesome. A speeding cosmic ray particle does not get very far into the upper air before it encounters a nucleus of nitrogen or oxygen, shattering it into a mess of debris that hammers yet more nuclei, whose debris smashes more, and so on, as a shower of particles the width of a sports stadium rains to the ground, the original cosmic ray long lost. You are sitting in such a shower, maybe more, right now. Ultimately included are neutrinos, gamma rays, electrons, neutrons, and strange atomic particles only rarely encountered. Cosmic rays gave us the first "atom smashers," from which we had our first glimpse of "pions," particles of the strong force that holds nuclei together, "muons" that are heavier versions of electrons, and "positrons," positive electrons that gave us our first view of antimatter. As neutrino generators, cosmic rays aided in the first actual determinations of their tiny masses, which were initially asserted from solar observations, so much of science so surprisingly converging together.

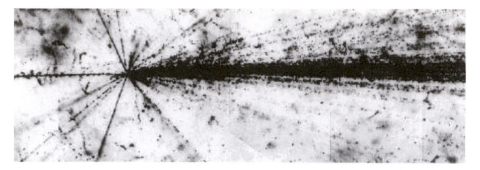

Figure 7.1. A speeding cosmic ray collides with an atomic nucleus in a detector's emulsion, shattering it into dozens of pieces that go flying away, leaving their tracks behind them. A cosmic ray hitting an atom in the Earth's upper atmosphere would produce a similar effect, and a shower on the ground. NASA.

These particle "air showers" were first detected by scientists studying the ionizing effects of radiation from uranium, radium, etc. The researchers found—quite by accident—that some ionization continued even when the radioactive materials were put away. In 1912, Vienna's Victor Hess (1883–1964) took an ionization detector on a balloon flight to a then-astonishing 17,500 feet and found the ionization effect to increase with altitude. Some mysterious radiation was coming in from space—*cosmic* radiation. By 1929 we had found the real source. Cosmic rays are not electromagnetic radiation, but high-speed ions that leave detectable tracks as they slam through various detectors. Satellite observation from completely outside our blanket of air gives us their count and chemical composition, which in turn tells us something about where they come from.

In parallel, the radioactivity studies that led to the accidental discovery of cosmic rays owe themselves to a yet-earlier accident. In 1896, only 16 years before Hess took his daring flight, Antoine Becquerel (1852–1908) was looking for fluorescence (a benign glow-in-the-dark caused by prior exposure to light) from a rock that contained uranium. He unwittingly discovered that the mineral exposed photographic plates right through the packages in which they had been wrapped. The rock was releasing penetrating radiation on its own. It was "radioactive." Along with the Curies (who were instrumental in further defining radioactivity), Becquerel won the Nobel Prize in 1903. Hess got it in 1936 for his discovery of cosmic rays, Carl Anderson for the discovery of the positron.

"Energy, People, Energy!"

As heard by generations of dance students and athletes. Energy. A term loosely used, one intuitively understood, one that in physics is precisely defined. Energy comes in a variety of forms—heat, motion, potential, electrical, magnetic—all of which can transform back and forth into one another. The most familiar form is that of motion, "kinetic" energy being the ability of one body to impart a force to another to change its motion, to impart heat, etc.

In ordinary life, the energy of a moving body is one-half of its mass (in kilograms) times the square of the velocity, a relation greatly under-appreciated by speeding drivers. But only if you do not go too fast. As you approach the speed of light, the rules of Einstein's relativity take over, and the relation becomes much more complex. Einstein showed that as a body gets closer and closer to light-speed, both its mass and energy of motion relative to a stationary observer increase toward infinity. At "c," a physical body—even an atomic particle—would have infinite mass and energy, which is impossible. Only massless photons of light and other electromagnetic radiation can go at this maximum speed.

Energy is expressed by a variety of units. The standard is the "joule," named after the English physicist James Prescott Joule (1818–1889). It is most easily understood through the familiar "watt" (James Watt of steam fame, 1736–1819), a unit of power that corresponds to the energy release of one joule per second. A hundred-watt lightbulb radiates 100 joules in one second, one joule in 0.01 seconds. Cosmic ray people, on the other hand, love using the "electron volt," or eV, which is the energy imparted by accelerating an electron through an electrical potential energy increase of one volt. One joule equals roughly six billion billion (6,000,000,000,000,000,000) electron volts. While the two units are redundant, they are useful within different energy ranges, the joule for large energies, the electron volt for small. Among the most astonishing thing about cosmic rays is that their kinetic energies span *both of them.*

Kinds

"Cosmic ray" is a sort of catchall term for three different kinds with three different sources. At low energies, the Sun sprays out "solar cosmic ray" particles during flares and coronal mass ejections, which can have quite an effect on orbiting satellites. A minor version called "anomalous cosmic rays" is composed of ordinary atoms from interstellar space that are accelerated toward us through low-energy interactions with our shock-wave solar bubble 100 or so Astronomical Units out. Then there are the "real" cosmic rays, "Galactic cosmic rays," those of the Galaxy, of the large-scale cosmos, which come to us from the stars themselves.

While seeming exotic, Galactic cosmic rays—"CR" in the trade—are hardly uncommon. Quite the opposite. At the top of the Earth's protective atmosphere, each square meter gets hit at the rate of thousands of particles per second. The rate and number-count peak at around 100 million electron volts, 100 "MeV," a tenth of a trillionth of a Joule. Below 100 MeV, CR are diminished as a result of the sweeping action of the outbound solar wind, which scoops them up and keeps them at bay. We have at this point no way of knowing how many of these low-energy Galactic CR there are, as we have yet to get outside the boundary of the wind at 100 AU, where it crashes against interstellar gas and the anomalous CR are born. We might, however, get the chance. Launched in 1976 to explore the outer planets, the *Voyager* spacecraft are beginning to penetrate the heliospheric boundary as they prepare to sail into true interstellar space. If the plutonium oxide power packs hold out (they are good until about 2020), we may finally get our first glimpse of the true low-energy CR rate. For now, we are literally in the dark about it (and them).

Now take a monster rocket ride toward ever-higher energies. As we climb above 100 MeV, the number count of incoming Galactic

CR drops like the proverbial bomb, while the energies rise to utterly astonishing levels. Roughly, for every increase of a factor of 10 in energy, there are 0.002 times fewer incoming particles with equal or higher energies. At energies of 100,000 MeV, we count only one particle per square meter per second rather than thousands. Up and up, higher and higher. At 10,000,000,000,000,000 MeV we get but one particle passing through our standard square meter per year, and just beyond 10,000,000,000,000,000,000 MeV we encounter one per square *kilometer* per year. But look at this energy. Forget electron volts: how about 1 joule, enough for a second's worth of light from a 1-watt bulb. The rarity of the highest-energy CR means that the upper limit is unknown. The highest ever recorded is around 50 joules: the kinetic energy of a baseball pitched at 35 miles per hour. All from a *single atomic particle no bigger than 0.00000000000001 centimeter across.*

At low energies, where there are lots of cosmic rays, direct particle detectors can be flown in high-altitude balloons and satellites. (The Hubble Space Telescope, an inadvertent detector, is bombarded by them. They produce false signals on the telescope's detectors that look like stars or other celestial objects, and must be expunged before images can be used.) But at higher energies you would wait a long time to get a hit, if you get any at all. You can't fly a square kilometer detector! At middling energies, CR people make use of the showers of collisional debris that hit the ground (which are easily detectable). At the highest energies, where even collisional showers are exceedingly rare, you can use the wide open sky, as the speeding atomic wreckage creates detectable flashes of light. At the very end of the observed energy scale, rarity wins, and we see none at all. Do CR exceed that 50-joule limit? We have no idea. But it's rather a scary thought that they might.

Given the number of CR that hit our tiny Earth, think of the vast number that must roam the Galaxy. The total amount of energy tied up in cosmic rays is roughly equal to that held by all starlight: the electromagnetic radiation from 200 billion stars. That's about half of the Galaxy's total energy. The other half is divided by the Galaxy's magnetic field and by the turbulent motions of interstellar matter. The particles may be small, but collectively, cosmic rays are a potent part of nature that must indeed be reckoned with.

Composition

Neither stars, planets, nor cosmic rays can be understood without knowing what they are made of. Like the Sun, CR are mostly simple protons—hydrogen nuclei. Next in order, about 10 percent of the total, are alpha particles, helium atoms that are stripped of their electrons. The small minority, a tenth of a percent or so, is made of all the rest of the chemical elements, the ratios more or less following what we find in the Sun and other stars. Atomic nuclei up to and heavier than iron have been found, and while the real heavyweights have not been directly detected, we are about as sure as we can be that all the chemical elements are present. In addition to these nuclei, there is a constant flux of electrons as well as their antimatter cousins, positrons (electrons with positive charges).

The curiousness of cosmic rays lies not in their resemblance to solar abundances, but in the detailed differences. The greatest of these involve the three light elements lithium, beryllium, and boron. Falling between helium and carbon, they respectively carry 3, 4, and 5 protons. In the Sun and in the cosmos in general, the three are exceedingly rare. For every million hydrogen atoms, there are between 100 and 1000 of oxygen, carbon, and nitrogen, while cosmically there are but 0.002 of lithium, 0.006 of boron, even less of beryllium. Cosmic rays have 100,000 times as much, clearly telling some kind of story. Large, though more modest, enhancements are seen for fluorine and for metals just lighter than iron, including titanium, vanadium, chromium, and manganese.

Long-Lost Origins

Taken together, CR energies and chemistry begin to tell us where these high-speed particles come from, and what happens to them on their flights to Earth. They do not give away their secrets easily. Many, indeed, are still tightly held. Starlight comes straight at us and thereby immediately reveals its source. You just look and there is the star. It's so obvious, it's seemingly silly even to discuss. Celestial radio waves, X-rays, and the like are similar. We merely need to look and we see their origins, or can identify them with some optical source. (Not that all are so identified; many are not.)

All we need do then is to note the direction of the incoming cosmic ray, and look to see where it came from, perhaps from some optically visible object. But now the obvious no longer works. Electrically charged particles are deflected by magnetic fields from their natural straight-line motions. Earth is surrounded by particles from the solar wind. Trapped within the Van Allen belts by our planet's magnetism, they circulate back and forth, their directions of origin long lost. Magnetic fields are everywhere. The Galaxy's is caused by mass motions of electrically charged ions as the system differentially rotates (the rotation period depending on distance from the Galaxy's center). Though more or less aligned along the disk and spiral arms, internal motions of gas clouds have made quite a tangled magnetic mess. Caught in the Galactic magnetic web, cosmic ray particles are continuously accelerated into long, curving paths. Most gradually get to the Galactic edge where the magnetic field weakens, and then escape the Galaxy. A tenth are destroyed when they smash into ambient interstellar atoms. The tiniest fraction slam into Earth. By then the flight directions as they hit us have been quite randomized. With the exceptions of those of the very highest energies (to be brought up later), there is no direct way to know where the things come from.

As in life, however, being indirect can be equally effective. The high energies of cosmic rays demand high-energy sources. And little that we know of is more powerful than a supernova, a whole star that explodes as a thermonuclear bomb. That stars can blow up certainly gives one thought. The Sun, after all, is a star. Will it be there for us tomorrow? A half-century of study on how stars live and die tells us not to worry. It's only the rare high-mass ones that create these majestic celestial fireworks. And the vast majority of cosmic rays.

Figure 7.2. NGC 3370, a striking galaxy in the constellation Leo very much like our own, displays a beautiful set of spiral arms packed with clumps of hot, young stars that ionize nearby gas clouds from which the stars were born. Rotation and local movements create a tangled galactic magnetic field that holds about a quarter of the galaxy's energy. Though not visible, the huge system is filled with the accelerated, curved paths of high-energy cosmic rays that are created by exploding stars and that hold yet another quarter of the energy. The Hubble Heritage Team and A. Riess (STScI).

To know supernovae, however, we have to look into the stellar aging process, at what is rather erroneously called "stellar evolution."

Star Lives

Like anything else, stars are born, live, and—as the energy sources that support them give out—die. Single stars have two paths to their final ends that depend on their birth masses. (Doubles, which are very common, are part of a separate story to be told later.) Following a quick birth from the dusty gases of interstellar space, all stars—historically called "dwarfs" to set them apart from dying stars called "giants"—run off the conversion of their core hydrogen into helium either by direct fusion of protons (as in the Sun) or by using carbon as a catalytic helper (as in the more massive stars). The result, though, is the same.

The Sun, an ordinary dwarf, initially had enough internal hydrogen to run stably (spitting out its neutrinos) for 10 billion years. Higher mass stars have more available fuel. But they are so much hotter inside that the rates at which they burn it more than offset the size of the hydrogen gas tank. As a result, the higher the star's mass, the *shorter* its hydrogen-fusing life. At the high end, more than 100 solar masses, while stars can burn with the light of millions of Suns, they live but for a few million years. Fortunately, such monsters are very rare, as you don't want one hanging around close to you. At the other extreme, stars with masses much under 90 percent solar last longer than the current age of the Universe. Here is where we find the vast majority of stars, dim little bulbs down to 8 percent solar, at which point hydrogen fusion as we know it stops. None of these lightweights has ever died! Substellar bodies below the fusion limit, called "brown dwarfs," abound too. They are so dim we do not know how many there are or what the bottom mass limit really is—their masses may overlap those of planets.

For all the complexity of the subject, it simplifies to a line drawn in the sand at about 10 solar masses. It could be 8, maybe 12, but this is astronomy, where factors of two sometimes don't much count, so we'll take 10. Start with the below-10 crowd. Given nothing else, stars would contract under the force of their own gravity, a process that would still generate considerable heat and light, but not for very long. The Sun's hydrogen-fusing core is prevented from doing so by the energy generated by nuclear fusion. Except to maintain conditions, gravity is not for now needed. When the core hydrogen runs out (10 billion years for the Sun, 20 million years at 10 solar), the core *has* to contract and become denser. But contraction also makes the core heat up, which fires up additional hydrogen in a shell around the quiet helium-rich center. The

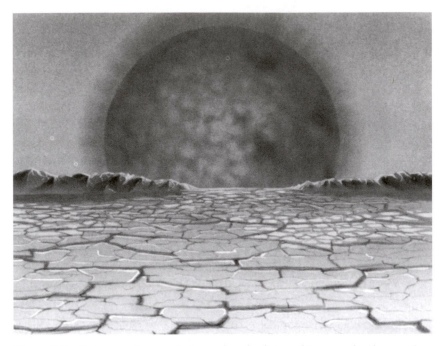

Figure 7.3. Rising as a huge red giant a hundred times bigger and a thousand times more luminous than it is today, our future aging Sun will someday blister Earth. At the end, it will have shrunk to such a small size that it will appear as no more than a brilliant star in a dimmed daytime sky. From *Stars*, Scientific American Library, New York: Freeman, 1992, © J. B. Kaler.

additional gravitational and nuclear energies—which can be only temporary—make the rest of the star *expand* and glow much more brightly. The expansion, though, also causes the surface to cool, and the star starts to become a "red giant."

Eventually, the dead core reaches a critical temperature of 100 million degrees, so hot that the helium, the ash of the old reaction, becomes a fuel and fuses to carbon and oxygen. At this point, the Sun will be 1000 times brighter than now. You can guess where that puts the Earth. The new energy source stabilizes the core for a time at somewhat lower luminosity, preventing further shrinkage. You can find such stars all over the sky, given away by their orange colors: Arcturus, Aldebaran, the list goes on. Go out and look for them.

Helium though, does not have the nuclear energy capacity that hydrogen does, so "helium-burning" does not last very long. When the core is fully carbon-and-oxygen-ized, it contracts *again*, which expands the outer parts of the star even more. At the peak of the action, a cool red Sun will be thousands of times brighter than now and will expand to the size of the Earth's orbit, not doing us any good.

But now the unexpected. As luminosity and size go up, the gravity at the stellar surface goes *down*, and the growing giant begins to lose mass through a wind that blows at an ever-greater rate. At its luminosity peak, an extreme giant may be losing matter a billion times faster than the rate in the solar wind. The powerful wind may be good news for Earth. As the Sun whittles itself down, its gravitational grip on us will lessen. If the wind's timing is right, the Earth will spiral away and be saved from destruction. But it's probably good-bye to Venus, and most certainly to Mercury.

Within only a few hundred thousand years, the entire envelope of the star, the part that surrounds the hot, nuclear-burning zone will be sloughed away and sent back into interstellar space from which it came. All that now remains is a cooling carbon-oxygen core of *much* lower mass. With an average density of a ton in a cubic centimeter (a sugar cube), the dense leftover is stabilized when its electrons just can't get any closer, its only destiny to cool as an Earth-sized "white dwarf." (Not to be confused with an "ordinary dwarf" like the Sun and the rest of the hydrogen-fusing gang. In the logic of astronomy, the name derives from the colors of the first ones found. Young hotter ones are blue, old cooled-off ones red, giving us blue and red white dwarfs.)

As its ultimate white dwarf grows in the middle, the Sun will eventually lose half of itself back to space, while a star that starts out at 10 solar masses will lose over 80 percent of its initial mass to make a white dwarf of 1.4 solar masses, the famed "white dwarf limit" (the "Chandrasekhar limit," after Subramanyan Chandrasekhar, 1910–1995, and a matter for the next chapter). Such stellar cinders, the products of billions of years of stellar life and death, are all around us. You can't build a heavier white dwarf. If a star tries, it goes

Boom.

Stars that begin life with masses greater than 10 times solar will go through the same processes, but during helium fusion (into carbon) will grow so large—as big as the orbit of Jupiter if not larger—that they are called "supergiants." Orion's reddish Betelgeuse (the constellation's upper left-hand bright star) and Scorpius's Antares quite qualify. These gigantic stars eventually get so dense and hot inside that their internal nuclear fusion can go much farther, working its way toward heavier and heavier elements. Once it forms from helium fusion and contracts enough, a carbon and oxygen core will fuse into neon (not possible in the Sun), which in turn fuses into magnesium, which eventually fuses to silicon. Each fusion cycle taking vastly less time; in just a couple weeks silicon turns to iron.

Which goes to nothing else. Iron is a curious element. Nuclear energy ultimately comes from the tightening together of nuclear particles, of protons and electrons. Iron's protons and neutrons are packed so tightly that further fusion can squeeze no more energy from them. To build heavier nuclei instead *requires* energy. So as far as energy production is concerned, it's the end of the line. The star has actually grown an iron white dwarf buried deeply inside a supergiant husk that is trying as hard as it can to lose mass through an amazingly strong wind. The iron ball, however, is near or above the 1.4 solar-mass limit. The internal heat and crushing forces within such a body are so high that even the electrons freed from their nuclei cannot support it. Teetering on the brink, the iron core starts contracting, then—with nothing to hold it back—catastrophically collapses, becoming so dense that the iron breaks down and all the shattered debris just compresses into a ball of neutrons.

When the ball gets to be the size of Manhattan (the supergiant itself as big as a planetary orbit), the neutrons become so packed that they suddenly stop the collapse. Squeezing to a density even greater than that of an atomic nucleus, the sugar cube now stuffed with 100 million tons of neutrons, the ball violently rebounds. The resulting outbound shock wave, aided by a spectacular flood of neutrinos to which the hugely dense gases are now opaque, and violent turbulent acoustic waves (all the processes yet to be clearly identified), speeds through the surrounding stellar envelope. Heating it to a hundred billion degrees, the shock and the resulting intense nuclear reactions explode the star, releasing 10,000,000,000,000,000,000,000,000,000,000,000,000,000,000 joules. Even though 99 percent of this energy flies out in the form of neutrinos, the exploding star is still so intrinsically radiant—a billion times more luminous than the Sun—that it can be seen across much of the visible Universe.

The maelstrom not only creates nearly all the chemical elements, including a tenth of a solar mass of iron, but ejects them as well. As we will see, white dwarfs in double star systems can explode as well. Between the two kinds of supernovae and the processes that take place in lesser giants, all of which eject chemical elements made by nuclear processes, we can account for the existence of everything from carbon through uranium, and in pretty much the observed amounts.

In the year 1054, Chinese scholars recorded the appearance of a "guest star" in our constellation Taurus. Even though two thousand light-years away, as seen from Earth it rivaled Venus, and was so bright it could be found in the daytime. In the late 1700s, the French astronomer Charles Messier (1730–1817) compiled a list of about a hundred "fuzzy things" seen through telescopes (the "Messier Catalogue" also containing some obvious naked-eye objects). Number one on his list, M 1, now known as the Crab Nebula, lies right at the place the Chinese found their Guest. Currently expanding at 1500 kilometers per second, the Crab is now more than nine light-years across, twice the distance between us and the nearest star, Alpha Centauri.

At the center is the surviving extraordinary neutron star that

1. holds well over a solar mass;
2. is stuffed into a volume the size of a small town;
3. spins 30 times per second;
4. generates a magnetic field more than 1,000,000,000,000 times stronger than Earth's;
5. beams energy outward along a magnetic axis strongly tilted against its rotation axis; and
6. blows a violent wind with speeds near that of light that illuminates much of the nebula.

Figure 7.4. The Crab Nebula, the exploded remains of the great supernova of the year 1054, accelerates local ions to create some of the vast number of cosmic rays that circulate through the Galaxy. At dead center is a 20-kilometer-wide neutron star (not visible in this image) that powers the nebula. Hubble image: NASA, ESA, J. Hester, and A. Loll (Arizona State University).

Acting like a celestial lighthouse, it sends us powerful bursts of energy every 1/30 of a second all along the electromagnetic spectrum, from gamma rays to radio. From our perspective, the star literally flickers on 30 times per second, and in between bursts (when the rotating beam is pointing away from us) is totally dark.

More than 100 expanding remnants of past explosions are known throughout the Galaxy, some with visible neutron stars, others without. The neutron star count is around 1500, so there must be many more that are hidden or even dissipated. While the Galaxy gets maybe one of these "core-collapse" supernovae a century, their effects last for millennia before they finally dissipate their energies into space. With the exception of the youngest, most of these explosive remnants are made not so much of the stellar debris, but of interstellar stuff that is swept up by the violent shock wave, which can heat the surrounding gas into the hundreds of thousands of degrees, enough to radiate across the electromagnetic spectrum, even unto X-rays and gamma rays. Whether core-collapse or white dwarf explosion, the energetics of the explosion and resulting shock waves are plenty enough to power the high energies and speeds of cosmic rays. Moreover, some of the remnants' radiation is clearly coming from electrons propelled at near the speed of light that are trapped in the remnants' tangled magnetic fields. Almost by default, supernovae are one source of cosmic rays.

Supernova debris has its own distinctive chemistry that comes from the nuclear explosion going on inside the expanding stellar envelope. Cosmic rays, on the other hand, for all their peculiarities, have compositions that far more match the chemistry of the Sun, which tells us that we are not directly getting hit by the supernova debris itself, but by interstellar atoms that are being accelerated by the outbound shock waves. The process starts when the shock hits an atom, gives it energy, and ionizes it if it was not already. The omnipresent tangled magnetic fields both in front of and in back of the shock then have a chance to keep a charged particle bouncing back and forth across the violently moving shock, each passage giving the particle more and more speed until it finally escapes. Deflected by the Galaxy's magnetism, it gyrates into huge loops tens of thousands of light-years across until it leaves the Galaxy or perhaps smashes into Earth or some other body, and we count yet another cosmic ray. Some CR, an unknown fraction, may even be coming in from far more distant supernovae that have gone off in other galaxies, these cosmic runaways giving us a direct touch with the Universe at large.

We seem to have solved the problem, except for

That Pesky High End.

Supernova blast waves are big and powerful enough to explain cosmic rays with energies that go to about a thousandth of a joule, at which point the numbers are *way* down, billionths of what they are at the peak. And we still have a factor of some 100,000 to go to get to the top of the energy pile! Where do those extra-energetic particles come from? Candidates have included especially high-energy supernovae (of kinds that we will encounter later), electric fields generated by the collapsed neutron-star pulsars, acceleration by shock waves set up by colliding galaxies, even by colliding clusters of galaxies.

The solution comes from looking at the rarest cosmic rays, those at the very highest end. At energies above about a joule, the Galaxy's magnetic field has insufficient strength to contain the spiraling particles at all, and they leave it on more or less straight lines with little chance of encountering anything. Given that they quickly leave *our* Galaxy, they must leave others as well. We should then be able to trace them back to their points of origin. But there are so *many* galaxies out there, so many points of possible origin, that it becomes quite impossible to make a match.

But there is an out. In the beginning—just after the Big Bang—the Universe was filled with gamma rays. As the Universe expanded and cooled, part of the energy was turned to mass: the protons, neutrons, and electrons of today. Over the past 14 billion years, continued expansion has cooled the leftover radiation bath to just 3 degrees above absolute zero, and we are left with the radio-wave "Cosmic Microwave Background."

Electromagnetic radiation's apparent wavelength, hence energy, depends on the observer's movement relative to the source. At the highest energies, now above 10 joules, cosmic rays move through this low-energy cosmic radio bath with such oncoming speed that the benign radio photons are Doppler-shifted to ultra-short wavelengths, and are thus felt by the CR as if they were gamma rays. Collisions with them smash the super-particles into scattered subatomic bits much as the CR are destroyed when they hit the Earth's atmosphere. There are so many background photons, that so far as the cosmic rays are concerned, they are running through a mine field. They thus cannot get very far, and therefore must be coming from relatively nearby where there are far fewer sources to confuse things. These highest-energy cosmic rays may be coming directly from the cores of "active galaxies," where central black holes (discussed in chapter 9) with masses hundreds of millions of times that of the Sun propel polar flows at near the speed of light. The rest of the high-energy gang above the accepted supernova cutoff are probably coming from the vast numbers of others at far greater distances, in addition to the various other possible sources mentioned above. We are being contacted not just by stars but by galaxies at the farthest reaches of the Universe!

Origins aside, without cosmic rays in all their forms and energies, life would not only not be the same, it wouldn't be at all. There might even never have been a Sun or a star-filled sky. There are half a dozen (if not more) good reasons to pay attention to cosmic rays, some quite surprising. Start with the most "outer" of them.

(1, 2) Star Formation

Remember how scientists once "proved" that bumblebees can't fly? But the bees didn't read the papers. Stars can't form either. The dusty blobs of interstellar matter from which they descend are all to some degree rotating, the result of cloud-cloud collisions, gravitational interactions, and general turbulence. In the absence of outside forces, the angular momentum of a system is constant (explaining why chapter 2's Moon is moving away from us as a result of Earth's slowing rotation). The initial clouds are huge, light-years or more across. Contract one to the size of a star and its spin would increase so much that it would fly apart. Therefore we don't exist.

Yet we do. So something must slow the cloud-spins. But how do you control something that may weigh as much as or even more than the Sun? With energy and time. Dust makes a cloud impenetrable to heating starlight, rendering it dark and cold inside, the chill a necessary condition for contraction to stars. Its atoms and

molecules would thus be neutral, un-ionized. Since low-level magnetism affects only charged particles, the Galaxy's magnetic field
threading through such a neutral black blob would have no effect.
Cosmic rays, however, are not easily stopped. Invading the cloud,
they leave long trails of ions behind them. The electrified ions grab
onto the magnetic field, which then resists their collective rotation.
Collisional coupling between the ions and neutrals then applies
the brakes to the whole system. Contraction eventually causes the
cloud to slip its magnetic noose, but by then it's rotating slowly
enough for the star actually to form without spinning itself apart.
The residual rotation is just enough to create a dusty disk, from
which cometh planets, comets, asteroids, and the resulting collisions with Earth described in the previous two chapters and that
have helped control life over the aeons.

Ion creation by cosmic rays also helps drive a cloud chemistry
that can produce complex organic molecules (number two in our
list). Included in the final mix are water, ammonia, formaldehyde,
various esters, ketones, cyanides, acetylenes, weird unstable strings,
some 150 kinds, and surely many more yet to be discovered. Per-

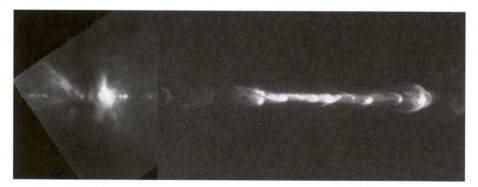

Fig 7.5. A powerful jet emerges from a dusty, dark, bright-edged, disk that hides
a brand new star. The star and circumstellar disk formed from a contracting
cloud of interstellar matter whose rotation was initially slowed by the Galaxy's
magnetic field, compliments of cosmic rays, which also helped create complex
molecules. Planets may even now be forming within the disk. Without cosmic
rays, they would never exist. Nor would we. Hubble image: NASA, A. Watson
(Universidad Nacional Autónoma de México), K. Stapelfeldt (JPL), J. Krist
(STScI), and C. Burrows (ESA/STScI).

haps the very seeds of life are out there to be brought in through the collapsing cloud that made the star that made the Earth that made ourselves.

Supernovae produce the bulk of cosmic rays and stir up the interstellar medium, while starbirth (hence starlight) is controlled by chaotic interstellar cloud motions, cosmic rays, and the Galactic magnetic field. No wonder the set of four Galactic energy reservoirs all contain about the same energies, since everything is connected to everything else. Even then to us. Even to our wash.

(3) Laundry

Out in the far western United States are areas rich in borates, chemicals made from boron and oxygen sometimes coupled with other elements. Though boron is cosmically quite rare, various geological processes have concentrated its salts into places like the Bonneville Salt Flats and Death Valley, where they have long been extracted, and were once drawn out by the legendary 20-mule teams. The stuff is best known as "borax," which is a prime ingredient for laundry and hand soaps. Boron (B) is used for lots of other things too, as are its even rarer sister elements, lithium (Li) and beryllium (Be), these elements with weights between helium and carbon often lumped together as "LiBeB."

The light three have no place in stellar element factories. Ordinary evolving giant stars do not make them, and neither do exploding supernovae. Stellar hydrogen fuses to helium, but then helium fuses directly to carbon, causing the LiBeB trio to be skipped over. Rather than manufacturing them, stars instead do just the opposite. The triplet is easily destroyed in stars by nuclear reactions that run at relatively low temperatures, only a couple million degrees, as convection brings surface gases downward. One way of estimating stellar ages is to look at the stars' lithium contents. Young stars have a lot, taken from their forming interstellar clouds, while the ancient Sun, having gotten rid of most of it, has hardly any at all.

Even though rare, astronomers still have to explain where these light atoms come from. The key lies in their much higher abundances in cosmic rays. There must be a connection. While the atoms of general interstellar space are rare too, averaging about one per cubic centimeter (a vacuum that would thrill a physicist), they are pervasive over a huge volume, that of the Galaxy itself, the Galaxy that is the CR racecourse. Given time, a speeding particle may collide head-on with an ambient atomic nucleus. Ninety percent of interstellar matter is hydrogen, simple protons, which are impervious to damage. But if a cosmic ray hits a heavier nucleus, it can break it apart into lighter components, a process called "spallation." Specially favored for destruction because of their relatively high abundances are carbon, nitrogen, and oxygen, whose byproducts include the little three: lithium, beryllium, and boron! Spallation of interstellar iron and nickel helps make metals like titanium and manganese that are richer in CR than they are on Earth, the collisions smoothing out a jagged solar-terrestrial graph of element-quantity versus proton number.

Almost all of these three light elements are products of CR collisions that made their way to Earth through the collapse of an interstellar cloud. Think about that next time you wash clothes. During which time you might pick up a science magazine and read about the latest discoveries in archaeology.

(4) From Lost Times

Archaeology, the science of the ancient, of those who lived in the far-distant past, requires a knowledge of *how* far in the distant past. One of the great needs of archaeology is the dating of prehistorical peoples, of their events and their artifacts, a difficult task made easier through the atomic transformations brought about by natural radioactivity.

The key in any radioactive dating method is to find an unstable isotope that exists in some reasonable abundance within the object of attention (be it plant or planet) and that has a decay rate comparable to the object's age. The Earth and Solar System can in part be dated to 4.6 billion years by comparing the amount of lead-206 (the daughter) to uranium-238 (the parent), which has a "half-life" (the time it takes for any sample to cut itself in half) of 4.5 billion years. There is therefore enough of both parent and daughter to measure. Uranium-dating will not work with things on archeological timescales, which are measured only in tens of thousands of years. Even if ancient remains incorporate uranium (some do, some don't), the decay rate is far too long: 50,000 years would make no significant difference in the initial amount.

Radioactive carbon, however, is another matter. Carbon, the basis of life, has three significant isotopes. Ninety-nine percent of all carbon is totally stable C-12, with six protons and six neutrons. The remaining 1 percent, with an extra neutron, is equally stable C-13. Within a tiny leftover is unstable C-14 with two extra neutrons, which has a half-life of only 5730 years. (No other carbon isotopes have any importance: C-11 decays in 11 minutes, C-10 in 19 seconds, and it's rapidly downhill from there.) With its short half-life, not that much longer than radium's 1600 years, a large quantity of C-14 by itself would be dangerous. But there is so little of it, one atom per trillion C-12 atoms, that it has no deleterious effects.

The question is why there is any at all. With such a short decay span, it should all be gone from the Earth and its air. Yet there it is. For C-14 to exist, it must be manufactured at such a rate as to balance its inevitable natural loss by steady decay. The factory is the same one that makes interstellar boron: cosmic rays. Among the secondary debris in a cosmic-ray air shower (caused when a speeding bullet of a particle hits an atom high in the atmosphere) are high-energy neutrons. If a neutron hits an abundant nitrogen-14 nucleus, it can kick out one of atoms's protons, sneakily replacing it with itself. The weight of the atom—hence isotope number—stays the same, but the proton number goes down by one. N-14 is thus turned to C-14. Over the whole Earth, the process makes about 20 grams of C-14 a day, which gets distributed around the globe and into the lower atmosphere by air currents, then ultimately bonds to oxygen to create radioactive carbon dioxide.

Any sample of C-14 will immediately begin to decay when one of its neutrons becomes agitated enough to spit out an electron. Losing a negative charge transforms the neutron into a positive proton, which turns the C-14 right back into good old N-14. Since electrons are nicknamed "beta particles," the process is called "beta decay," and is a powerful agent in the creation of chemical elements in stars.

The amount of C-14 at any one time will depend on the formation rate relative to the destruction rate. But we do not need to know the formation rate. All we need measure is the current amount and the decay rate for the archaeologists to get to work. We just have to believe in relative constancy of formation. Living things, made of carbon, all contain the ratio of C-14 to C-12 that is in the air or ocean during their lifetimes. When they die and no more carbon is incorporated, the C-14 dies away as well. By measuring the C-14/C-12 ratio in anything that was once, or once contained, organic matter, we can tell how old it is, up to around ten times the half-life, or about 50,000 years. Campfires and potsherds and other ancient relics of life thus begin to give away their secrets.

But

There is this vexing business of constancy. The cosmic ray flux on the Earth is not really steady, and thus neither is the C-14 formation rate. We can't automatically assume it's the same now as it was in prehistoric times, or even last month. While the actual Galactic cosmic ray flux outside the Solar System is dependent on the rate of historic supernovae, it's unlikely to change much over only a few thousand years. A prime culprit is the Sun. Lower energy cosmic rays are kept away from us by the solar and magnetic activity, and thus show an 11-year cycle as well as much longer-period variations that are not directly observable (recall chapter 3's Maunder minimum).

But we can still measure what it was. Take a core from an ancient tree. The long-lived bristlecone pine is usually the target. Measure C-14/C-12 in each of the growth rings, and we can calibrate the ratio against time (allowing of course for the C-14 decay rate). We can do the same with deep ice plugs taken from glaciers, in which the ice appears in annual seasonal layers. Turn the process around and it gives us a sense of historical solar activity.

Of course, our modern era messes it all up. Archaeologists of the distant future are going to have a real problem. Atomic testing during the late 1940s and early '50s sent a huge spike of C-14 into the air that is still being incorporated into all living things. On the other hand, fossil fuels (coal, oil) are so old that they have no C-14 at all. So as we warm the globe by burning them, we are going the opposite way, and are diluting the natural atmospheric C-14. The dating of the ruins of ancient Chicago 50,000 years from now will require some interesting calculations.

Through the medium of cosmic rays, distant supernovae thus allow us to know more of ancient life. But while important to our understanding of human life, of its development, C-14 is not *necessary* for it. Consider then a process that may be.

(5) Donner and Blitzen

Not reindeer: thunder and lightning. Lightning is, as we learn early on, a giant electrical spark (set of sparks really) that takes place within or between clouds or between clouds and the ground. The simple cause is separation of electrical charge, positive in one part of a cloud, negative in another. When the charge difference is sufficient between cloud locations or between cloud and ground, the insulating properties of the air break down and charge is exchanged to neutralize the difference. Lightning a million times more powerful than anything on Earth has been observed on Jupiter and Saturn; it is hypothesized on Venus and Mars, and is probably ubiquitous in the Universe.

The *detailed* causes of lightning, however, particularly the cause of charge separation, as well as of many other phenomena associated with it, remain mysterious. Falling raindrops or ice crystals are usually purported to be the charging battery, but the exact mechanism is elusive. Lightning creates X-rays, even gamma rays. Lightning bolts are also associated with powerful emissions called

Figure 7.6. A powerful storm hammers the sea. Clouds and lightning, a deep part of life on Earth, may in part be caused by cosmic ray interactions with the Earth's atmosphere. Courtesy of Bruce Kaler.

"blue jets" and "red sprites" that can jump *upward* to altitudes of 50 to 100 kilometers above the ground. Difficult to observe, their causes are unknown.

The discharge that begins the atmospheric breakdown and leads to the complicated cascade that creates lightning must start from somewhere. There is a growing feeling that lightning sequences, at least a portion of them, are triggered by ultra-high-speed electrons produced in air showers by cosmic ray hits. These electrons accelerate those in the atmosphere, which hit more electrons, and suddenly there is a cascade of them that begins the breakdown process. Even the clouds that cover Earth may at least in part have their origins in CR, as clusters of ions produced by them may act as nucleating agents for water condensation.

About 2000 people are hit by lightning each year, of which a third do not survive. Lightning is the chief cause of forest fires. At the same time, it creates natural fertilizer, and at least one theory depends on it to have started the chain of events that created the most ancient life from simple organic molecules. Clouds provide us with life's water ("... Cool water ...") and in complex ways help control Earth's temperature. The reach of supernovae is great indeed. The effects are even greater for those intrepid astronauts.

(6) Bound for Space

Cosmic rays of all kinds—from the solar CR launched by coronal mass ejections to the Galactic variety—are an underappreciated danger for space travelers. For all the impact that cosmic rays have on us, astronauts in near-Earth orbit (as in the Shuttle or the Space Station) are still more or less protected from them by Earth's magnetic field. But go beyond a few Earth-radii and the trapping Van Allen belts, and the danger becomes more real. Going to the Moon and back via *Apollo* seems to have worked (thanks to a quiet Sun between activity peaks), but long-term exposure to the particles and to X-rays created when cosmic rays hit the skin of the craft will be dangerous. Manned spacecraft bound for Mars will have to be carefully shielded.

Worse yet is when any potential astronaut gets beyond the heliosphere, into raw space, where the Galactic cosmic rays rule. Travel to the stars, by whatever means, may be deadly. Science fiction has spacecraft zipping through the cosmos at near the speed of light so as to get from one star to another as fast as possible. Imagine the power of the impact of a head-on collision with a cosmic ray also flying near light speed! The resulting radiation bath caused by the constant hammering would be lethal. (Plowing through the dusty gases of interstellar space and through the Cosmic Microwave Background won't do the crew any good either. Friction with interstellar matter would probably melt the craft.) Cosmic rays, really their parent exploding stars, could help set the limits on how fast and far we can actually go.

For all their power and extensive effects, cosmic rays remain the intermediaries to humanity of supernovae. But what about the supernovae themselves? Direct attacks by such explosions might be even more potent, even on the life that cosmic rays may have had a role in beginning in the first place.

Chapter 8

Super Star

Mountain midnight. February 1987. Ian Shelton looks at an image he has taken of the Large Magellanic Cloud, a nearby small satellite galaxy to our vastly larger Milky Way. A trained eye sees an extra-bright star where none was before. Return outside, into the night: there it is, glowing softly, the first supernova visible to the naked eye since the invention of the telescope.

Go back a day. From Australia, Robert McNaught sees the explosion in progress, watches it brighten. The day before McNaught's discovery, Supernova 1987a had been a quietly evolving blue supergiant, a swollen star of some 20 solar masses that long ago gave up central hydrogen fusion. Now it is gone, turned into a blast of energy that dwarfs anything in the known Universe except for varieties of its own kind. At about the time of McNaught's sighting, 11 neutrinos from the direction of the Large Cloud are picked up by the Japanese Kamiokande neutrino telescope, while 8 more are recorded in another detector beneath Lake Erie. The number is right, about that expected from a collapsing stellar core. At the moment of collapse, a thousand trillion pass through you, far more than come from the Sun, leaving without a trace.

Within 75 days of discovery, the supernova brightens to "third magnitude." Third magnitude stars, of modest brightness, are mostly nearby, a few tens or hundreds of light-years away. Not 170,000 light-years. So far as we know (there is no way of exploring prehistory), SN '87a is the most distant single star ever seen with the unaided human eye, testimony to raw stellar power. Imagine such a thing not 170,000 light-years off, but maybe 100, even 10 light-years away. Though still farther than the nearest star, the

Figure 8.1. Supernova 1987a (at lower right) went off within a rich section of
the Large Magellanic Cloud, a small nearby galaxy 170,000 light-years away
that orbits our own larger one. The huge Tarantula Nebula, a vast region of
hot stars and dusty interstellar gas, is at the upper left. Most of the stars in
the picture are luminous giants and supergiants, but they are still completely
overwhelmed by the immensity of the explosion. Anglo-Australian Observatory/
David Malin images.

effects would sear the Earth. Similar events are commonly seen in
nearby galaxies: and take place in our own as well.

To explore such stars, there are tales to be told. The first begins
over 2000 years ago.

Yesteryear

Return now not to 1987 but to the second century BC. Hipparchus
of Nicea had no telescope (had no idea of what one even was), but
did have two treasures: dark skies and a dry, clear climate. These

and an inventive mathematical mind gave us among the greatest of antiquity's astronomers, rivaled only by Aristarchus of Samos (310–230 BC), who thought the Earth went around the Sun, and tried to find the solar distance. From the island of Rhodes, where Hipparchus spent most of his life, he had a superb view of the heavens. With his own observations he discovered the precession of the equinoxes and measured the orbital size of the Moon in Earth-diameters. And got the right answer. Eratosthenes' earlier estimate of the size of the Earth, from how the Sun's angle from overhead varied with terrestrial position, then gave the lunar distance in common units.

Hipparchus also gave us our measure of stellar brightnesses that two millennia later would still be used in charting the flow of stellar evolution, including what would someday be known as supernovae. Starting us on a long road, he divided the sky's stars into six brightness categories now called "magnitudes," first magnitude the brightest, sixth the faintest that he could see. In our own times, we express magnitudes in strict mathematical terms that Hipparchus probably would have loved. On a sliding scale akin to audio's decibels, each magnitude division corresponds to a factor of 2.512 in visual brightness. That is, an average second magnitude star is just over 2.5 times brighter than one of third, and so on, making a "star of the first magnitude" (Hollywood owing a debt to Hipparchus as well) exactly 100 times brighter than one of the sixth (the actual definition). Measure precisely to a hundredth of a division, pin the system to Polaris at 2.0, mess around with calibrations for a century, expand it to different regions of the spectrum, and you have the modern system (systems, really).

On this simple scale, there are 22 first magnitude stars: Aldebaran, Deneb, Regulus, Antares. Those of the Big Dipper are mostly second (of which there are 70), the Dipper's faintest being SN 87a's third magnitude (of which there are over 200). The four bowl stars of the Little Dipper, hard to see from town, are second through fifth. Thousands of sixth magnitude stars speckled Shelton's dark mountain sky.

From the technical definition, the generic "first magnitude stars" cover quite a range. Eight are so bright they are forced into magnitude zero (Vega, Arcturus, Capella), while two go into minus

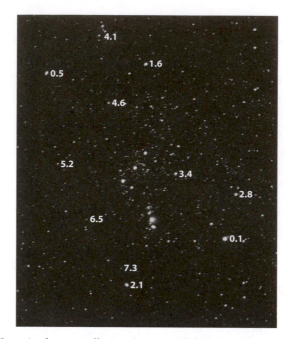

Figure 8.2. Stars in the constellation Orion are labeled to the immediate right with apparent magnitudes that range from Rigel at 0.1 and Betelgeuse at 0.5 down (in brightness) to 7.3, which is below naked-eye vision. The trio of stars in the center that make Orion's Belt are all second magnitude (from right to left, 2.2, 1.7, 1.6). J. B. Kaler, *Cambridge Encyclopedia of Stars*, New York: Cambridge University Press, 2006, reprinted with the permission of Cambridge University Press.

first, the top dog (Sirius, the "Dog Star") almost minus second. Then take on the planets. Lordly Jupiter is almost –3, while Venus beautifully hits as bright as –5, enough to be seen in broad daylight and to cast shadows at night. A quarter Moon lies at –10, the full Moon at close to –13, and going well over the danger zone is the Sun, near –27. In the other direction, in binoculars you might see to 8th or 9th, in a small telescope to 10th–12th, while in the largest and best of professional instruments, we image to 30th, a trillion times fainter than magnitude zero.

Apparent magnitudes, those as witnessed from Earth, provide limited information. To understand the stars we need true luminosities in watts, as we have for the Sun. The actual brightness of a star as seen from here depends on both wattage and distance: the farther away it is, the fainter it will look. Stellar distances are secured first through parallax. As Earth orbits, we see a nearby star from different positions in space, which causes it to appear to shift back and forth against the distant background. The farther the star, the smaller the shift.

The parallax shift creates a long, thin triangle whose narrow apex lies at the Sun. Once again, enter Hipparchus, who it seems also invented trigonometry. From the degree of shift and a bit of trig, we find the stellar distance in terms of the size of the Earth's orbit, in Astronomical Units, and thus (since we know the distance to the Sun) in any units we like, including light-years. The nearest star, Alpha Centauri, lies 4.4 light-years away. Parallaxes for hundreds of thousands of stars out to distances of 1000 light-years and more allow calibration of other methods that in turn allow us to walk the Universe.

Apply the distances to apparent magnitudes, and out come luminosities. But astronomers are simple folk. Instead of watts, it's easier to keep using magnitudes. From distances, we quickly calculate "absolute magnitudes," what the apparent magnitudes would be if, god-like, we could move all the stars to a standard distance of 32.616 light-years. (32.616? The professional unit of distance is the *parsec*, the distance at which the Earth's orbital radius would appear to be a second of arc, 1/3600 degree, across. The standard distance is a more logical 10 parsecs, which translates to the seemingly illogical 32.616 light-years.) Correction of the visually determined magnitudes for invisible ultraviolet light from hot stars, or infrared from cool stars, then leads to total absolute luminosities.

An absolute-magnitude sky would look very different. Not that far from 33 light-years, the northern "big three" (Vega, Arcturus, and Capella) would suffer only minor adjustments. Orion would gleam with several stars (Betelgeuse, those of the belt) as bright as Venus, while Deneb (at the top of the Northern Cross), glowing at −7 (with a total wattage 50,000 times that of the Sun), would rival a slim crescent Moon. Our blinding Sun, however, for all its 400 trillion trillion watt output, would fall to a mere fifth magnitude, roughly equaling the fainter stars of the Little Dipper. Only a few of the less luminous stars (by far the most popular kind) would be visible at all. At the top, the greatest of stars come in at an amazing −10 (a million times more visually luminous than the Sun), at the bottom they are down to a miserable +23, visually 16 million times fainter than solar. Ignoring infrared or ultraviolet, it would take 16 trillion of the faintest stars to equal one of the brightest.

The culprits in creating this vast divergence are mass and evolution, high-mass stars the most luminous, low-mass the least, coupled with brightening of lower-mass stars as they become more luminous giants. In the deep cellar, below the +23 limit, are substars, brown dwarfs that cannot initiate full fusion of hydrogen to helium. Many are so faint that they would not be visible to the eye at all, at any distance, even were they within the Solar System. The lowest limit is quite unknown. Stars of highest mass, those born with more than 10 Suns-worth of matter, develop collapsing iron cores, whence they become supernovae. SN 1987a hit a peak absolute magnitude of −16. Place the supernova and the Sun side-by-side, and the exploder would be over 200 million times brighter. That is but the start, as the range is very large. For reasons still unclear, SN '87a was underluminous. The average is around −18, over five times as energetic. The brightest one ever hit an astounding −22, 100 times as luminous as SN '87a.

But as Al Jolson sang 2000 years after Hipparchus, "You ain't heard nuthin' yet." Over 1000 times as much energy comes out in the supernova blast wave and in neutrinos. Moreover, there is another route to greatness, indeed to greater greatness, one that involves ordinary dwarfs that sneakily conspire to blow up white dwarfs. Within two decades of Jolson's 1919 song, all of it began to fall into place. But first, with due respect to Robbie Burns, we have to go back to

1054, 1572, 1885, and A' That.

While the Chinese Guest Star of 1054, which spawned the Crab Nebula as well as a host of cosmic rays, is justifiably famed, it was not the most important to the art of astronomy. That honor might

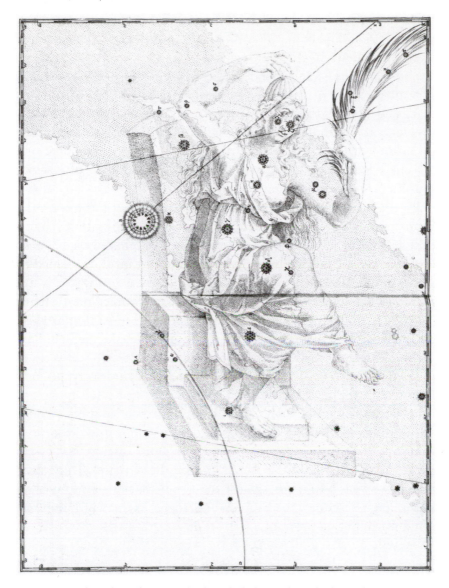

Figure 8.3. Though Tycho's Star had exploded into the naked-eye sky 30 years before, it was still included in Bayer's depiction of the constellation Cassiopeia in his great star atlas of 1602, the *Uranometria*, giving some sense of the brilliance and importance of the visitor that Tycho saw. Rare Book and Manuscript Library, University of Illinois.

better go to the "new star," the *nova* (Latin for "new"), that blossomed in Cassiopeia in 1572. As bright as Venus, it helped direct Tycho Brahe (1546–1601) toward his road of scientific stardom. Tycho was the last and by far the best of the naked-eye observers. His fame rests largely on precise observations of the positions of stars and planets that allowed Kepler to discover the basic laws of planetary motion. His laurels also stem from his careful observations of the fading of the nova, which forever after became "Tycho's Star" and which demonstrated, in defiance of dogma, that the heavens are not unchanging.

By the beginning of the twentieth century, lots of "novae" had been seen, many—about one every 20 years—hitting first magnitude or so. But Tycho's (and later, when we deciphered the Chinese and other texts, more of them) were so bright that they seemed to stand out from the rest of the gang, which as we now know are mere surface explosions on white dwarfs in double-star systems. The key to understanding the difference was the nova that in 1885 erupted to sixth magnitude within the Andromeda Nebula. The Nebula (Latin: "cloud") is an elongated, easily visible fuzzy patch in central Andromeda. At the time of the nova's discovery, nobody knew what the fuzz-ball was, though it was suspected to be another galaxy like our own. Around 1925, Edwin Hubble (1889–1953) resolved the white mass into individual stars and discovered that indeed it was a galaxy, one that rivals our own, and very far away (as we now know, some 2.5 million light-years). To be near naked-eye vision (though against the bright background invisible without a telescope), the erupter had to be extremely luminous, not a nova, but a SUPERnova.

I and II

Thank goodness for the Romans. Their numerals so nicely lend themselves to the naming of Important Things, from Stars to Superbowls. The era of scientific supernova study was truly birthed in 1941. Since supernovae in our own Galaxy are exceedingly rare, Walter Baade (1893–1960) and Rudolph Minkowski (1895–1976) had to study them in other galaxies, and saw from their spectra that they sharply divide into two groups. The spectra of ordinary stars contain "barcode" absorptions of the common chemical elements (each with its own set), from which we get all sorts of information, including their temperatures and chemical compositions. Under special circumstances, some of these absorptions can reverse themselves into emissions. Type II supernovae were seen to radiate strong emissions that are specific to hydrogen, while Type I had none at all (though emissions of other elements were present).

Further investigation revealed other distinctions. Larger galaxies come in two basic forms. In systems like ours, most of the stars are arranged in flat disks that contain considerable dusty gas from which *real* new stars are born. Star formation is continuous, and therefore the disks are filled with all kinds of stars, from ancient to brand new, from lowest to highest masses. *Elliptical* galaxies, on the other hand, have no disks, little or no interstellar stuff, no star formation. Since high-mass stars die young, ellipticals are empty of them.

Type II supernovae, those with spectral hydrogen emissions, are confined to galactic disks. Type I, however, are seen everywhere—in the disks, in the surrounding halos of disk galaxies (where there are no young stars either), and especially in elliptical galaxies. Since the disks are distinguished by massive stars, it does not take much deduction to figure out that Type II's derive from them, especially since basic calculations show that they develop iron cores that collapse to end their lives. The Crab explosion of 1054 is a prime example. But what about Type I? Massive stars are out. Type I's *have* to be produced by low-mass stars. But how? There is no way for the Sun to blow up. Something else must be at work, as was found by our next hero, Subrahmanyan Chandrasekhar.

Chandra's Limit

By the time Baade and Minkowski were divvying up supernovae, white dwarfs, the leavings of ordinary solar-type stars, were already well known. They are strange (from our limited perspective) critters. Within any star, the temperatures are so high that electrons become stripped from their parent atoms and act as a free gas. The ultra-high densities of white dwarfs are so great, however, that the electrons behave oddly. Only so many of them can be jammed into a designated cubic centimeter below a given velocity. There is no immediate limit to the number that you can cram into the sugar cube, but if you try, the new ones must go in at ever-higher speeds. It's a "quantum effect" that results from electrons behaving as much like waves as they do particles.

As the mass of a white dwarf increases, so do the internal pressure and density. In 1930, a young Indian astrophysicist named Subramanyan Chandrasekhar contemplated the matter by realizing that at sufficiently high mass, the internal electron-speeds must approach that of light. Such speeds distort the usual concepts of time and space, so Chandra was forced to invoke Einstein and relativity. He then found that at a critical limit, now know to be 1.4 solar masses, the white dwarf must catastrophically collapse and, as we now know, burn with nuclear violence. (Type IIs are variations on the theme. Supernova 1987a exploded when the star built an internal iron core of roughly a Chandra mass and found it could not sustain it.)

Ordinary white dwarfs, those created safely below the limit, are remarkably stable. Their only fate is to radiate away their internal energies, and, so far as we know, to cool forever. Place one in a double-star system, however, and the rules can dramatically change. Doubles—binary stars—are everywhere. Somewhere around half the stars in the Galaxy are members of such pairs, two stars (or more, making multiples) born together that spend part, even all, of their lives in orbit about each other. In some systems, the components are nearly identical, while in others, one can carry dozens of times the mass of the other. Component separations are equally diverse. At the one extreme, they are separated by hundreds, even thousands, of Astronomical Units, and orbit in tens of thousands to millions of years. At the other, they are in near (or actual) surface contact and whirl around in hours. The stars of wide binaries have little to do with one another, indeed one can hardly know the other is even there. In closely spaced doubles, however, each member can profoundly affect the other's aging processes.

In spite of being 30 Earth-diameters away, the Moon raises obvious tides in both the solid Earth and its waters. Earth raises even bigger ones in the Moon. Around the members of every orbiting pair are teardrop-shaped tidal surfaces called "Roche lobes" whose teary points connect to form a three-dimensional figure-eight. Along such a tidal surface, the gravities of a pair of bodies (combined with the forces of orbital revolution) neutralize each other, allowing a speck of matter placed on the surface to go anywhere. Earth and Moon—and more widely spaced stellar pairs—are seated neatly within their lobes and cause each other no particular harm.

The stellar aging process makes all the difference. If the two stars were born close enough together, when the more massive evolves first to become a giant, it can grow so much that it can fill its lobe. (The evolving star could also be a supergiant that explodes; here we just look at stars under the 10-solar-mass limit.) Mass can then escape altogether, or it can flow to the smaller, tide-inducing companion. Friction from the enveloping gas will cause the two to spiral toward each other. The white dwarf that results from the evolution of the more massive star, a dense ball of carbon and oxygen, now finds itself *really* close to its still-unevolved companion.

Now, the "tides turn." If the two stars have approached each other closely enough, the white dwarf can reverse the process. Fresh Matter rich in hydrogen now flows from the unevolved star onto the white dwarf. The huge gravity of the white dwarf compresses and heats the infalling layer. When thick and hot enough, the fresh hydrogen violently erupts, burning quickly to helium in a natural hydrogen bomb. The reactions are much the same as those that go on inside a star, but are uncontrolled. And on Earth we see a bright nova.

The white dwarf's interior is so dense that it is completely unharmed. After a few years or decades, long after the hydrogen layer is burned off, the white dwarf settles back to normal, and the whole process starts over again to create another nova perhaps 10,000 or 100,000 years down the line.

However, if the white dwarf is the progeny of a fairly massive star (one just below the supernova limit), it will have a mass close to the Chandra limit. Then everything happens faster, and the pop-off interval is greatly shortened from millennia to centuries, even decades, to create a "recurrent nova." Many such dot the sky, the best one blazing away every 20 years or so. When a nova explodes, there is a little residual mass left behind: after each eruption, the white dwarf weighs a bit more. For a low-mass white dwarf, the gain is not enough to be mentioned. But for a high-mass firecracker, the gain over the years may be sufficient to throw the white dwarf *over* the Chandra limit. It's now the end of the line as the whole star flames out in a 1.4 solar mass nuclear bomb, the resulting Type I supernova utterly dwarfing the nova events that preceded it.

Another, quite likely even better, route to oblivion might be the merger of two orbiting white dwarfs whose combined mass exceeds Chandra's limit and that have been brought very close together by the evolutionary processes described above. As the two revolve around each other, they cause swirling ripples, "gravity waves," to radiate outward into the fabric of Einstein's spacetime. Losing energy in the process, the two spiral inward, closer and closer, until they merge and destroy themselves.

Since the time of Baade and Minkowski, other kinds of supernovae with no hydrogen emissions have been discovered. Types Ib and c are core-collapse events like Type II whose parent stars had already lost their hydrogen exteriors through powerful winds. The original Type I white dwarf variety then became known as "Type Ia," of which Tycho's Star is the most famed example. Usually (though not always) far more powerful than the core-collapse variety, Ia's typically reach absolute magnitude –19.3, which amounts to nearly 5 billion Suns of visual energy, enough that at 32.6 light-years one would shine with the light of 480 full Moons. So bright that they can be seen over distances of billions of light-years, and so uniform (the full range only about a magnitude), they have become the chief means by which astronomers measure the distances to ultradistant galaxies, and thereby determine the expansion and acceleration of the Universe. Not only is the white dwarf (Ia) supernova ordinarily several times brighter than its typical lesser Type II (core collapse) brother, but in the ensuing nuclear maelstrom, on

Figure 8.4. A Type Ia, white dwarf, supernova erupts in a galaxy more than 6 billion light-years away. At its peak, the supernova seems to outshine the faint galaxy, whose center is to the upper left. NASA and A. Riess.

the average it ejects three-tenths of a solar mass of iron, thrice the yield of the core-collapsers!

March of the Supernovae

Penguins march from the Antarctic coast to their breeding grounds, an arduous journey of hundreds of miles. Now watch the naked-eye supernovae do the same, except in the opposite direction, from their breeding grounds toward *us*.

1. Begin, at least in principle, 2 million light-years away at the Andromeda Galaxy. While the 1885 supernova (known as "S Andromedae") was not visible without a telescope, it *would* have been if seen in a dark sky if there had been no luminous galaxy background behind it. Its absolute luminosity near −18.5 strongly suggests white dwarf collapse. At its position is now an opaque iron-rich cloud expanding into the cosmos and merging with the local interstellar gases.

2. Now take a giant step to 170,000 light-years (15 times closer) to Supernova 1987a. Though a lesser version of Type II core collapse, with a relatively faint absolute magnitude of –16, at visual third magnitude, comparable to the faintest star of the Big Dipper, it could easily be seen from the backyard of a modestly lit town. The brilliant light of the supernova first illuminated a huge hourglass-shaped shell that the evolving star had lost through winds long before it exploded. Following along behind the speed-of-light flash was a powerful blast wave that started off at 30,000 kilometers per second, then coasted until it too slammed into the outer shell, lighting it even further, the process going on yet today. So the lessons for Earth begin.

3. A large skip in space to our own Galaxy gives us "Kepler's Star." Observed by the master himself, Kepler's went off in 1604 deep in southern Ophiuchus where it was hard to see, and disappeared quickly into twilight. Distance estimates for any supernova in our own Galaxy are difficult to make and prone to serious error. The best shot is about 18,000 light-years, which gave Kepler's a luster that hit apparent magnitude –2 or –3, roughly equal to the brightness of Jupiter. Its absolute magnitude at peak therefore hit roughly –19. That, a residence well into the Galaxy's halo, and a lot of ejected iron strongly suggest Type Ia (white dwarf overflow).

4. Now for the one that started the whole thing, at least in Europe, Tycho's Star of 1571, which—bright as Venus—hit apparent magnitude –4. That's bright enough to be seen in daylight and to cast eerie shadows on a dark night. The best distance estimate is 11,500 light-years, 65 percent Kepler's. Clearly Type Ia, it too hit –19 absolute and was so famed that it found its way onto Johannes Bayer's great atlas of the sky, the *Uranometria* of 1602, even though by then the star had long since faded away. If only there had been telescopes! At this point, supernovae begin to get a bit scary. Never mind marching penguins. Think again about our horror film's zombies treading steadily forward toward their victims.

5. The champion of recorded history, the great supernova of the year 1006, shone at us from only 7000 light-years away. Records from so long ago are hard to decipher and arguable (and are thus

vigorously argued). Glowing in the Milky Way in Lupus, far to the south for the observers of the time, a good guess is –8, even –10, the latter as bright as a quarter Moon, so bright that you could find your way in the dark by starlight alone. Even at the lowest limit of visual brightness, even with no allowance for dimming by the Milky Way's interstellar dust, the thing hit close to absolute magnitude –20 (making it most likely a "Ia" event).

6. Just slightly closer, at 6500 light-years, lies the remains of the Crab, the nearest of historical times. Backing off a bit in brightness, it hit a "mere" visual magnitude of –4 or so, again rivaling Venus and again seen in daytime. At an absolute magnitude of about –16, the Guest Star of 1054 is clearly a core collapse Type II-er. Though no rival of the 1006 event, it was energetic enough to furnish some of our cosmic rays.

7. At 2000 or so light-years, the closest of the historical lot, the supernova of the year 185 is also the oldest, and therefore the most controversial. Good linkage with its remnant, though, has shown that the ancients really did see and record a –2 magnitude super-nova (of unknown kind) nearly two millennia ago.

Closer Yet?

Supernovae, coming from rare stars of greater than 10 solar masses, or from white dwarfs in scarce, precariously balanced doubles, are equally rare. Several lines of evidence suggest a current rate of be-tween one and three in the Galaxy per century. That the historical record shows only one every 250 years is largely the result of both the size of our Galaxy and the hiding power of thick interstellar dust. One of the nearest known, a core-collapse event just 70 per-cent farther than the Crab, was not seen at all. Its expanding rem-nant ("Cassiopeia A," one of the brightest sources of radio waves in the sky) suggests a date about a century after Tycho's, when astronomers would have been on the watch for them.

Still, even a couple of these "guest stars" per century is a very low rate. The odds of getting one near us are then small—though very real. There are several candidates for future close events. The best for the core collapse model are two red supergiants, Orion's Betelgeuse and Antares of Scorpius. At a distance of 640 light-years, the mass from luminosity (135,000 times that of the Sun) stands at around 20 solar masses, by far enough to put it over the line. At 600 light-years, Antares, carrying 15–18 solar, is similar.

It's impossible to say exactly what is going on inside such stars. The odds are that they are still fusing their internal helium into carbon and are still a million years from generating the necessary iron cores. But let's set them off tonight. Unlikely, but not impossible. Make them average. Betelgeuse's absolute magnitude of –6 coupled with distance gives it an apparent magnitude of 0.5, just on the line between zeroth to first. Kick the star up to the average absolute magnitude for core-collapse supernovae, 12 magnitudes brighter, and it would present a spectacular sight as bright as a gibbons Moon. You would not just "see things" but could nicely navigate at night. Antares would appear about the same. There would be no having to hunt for them in the daytime either; all you'd need to do is look up and there they would be, a sight nobody in recorded history has ever witnessed. (As long as we are imagining, let them explode at the same time. Roughly as one set, the other would rise, giving us nearly 24 hours of supernovae.) Either would be visible in the night sky for years as they ever-so-slowly faded away, destroying the outlines of their constellations as they disappeared.

One such event might have actually occurred. There is strong evidence for a supernova remnant (the exploded debris and resulting shock wave) some 24 degrees across in the obscure southern constellation Antlia. Such a size suggests closeness, perhaps as near as 200 light-years. Since it could not have occurred any earlier than a million years ago, only our pre-human ancestors would have witnessed a "new star" with a magnitude –14, three times brighter than the full Moon, though it would hardly have mattered, since they probably did not recognize any of the "old stars."

The white dwarf variety of supernova is even more spectacular. The march toward us is longer, as the nearest candidate recurrent nova is much farther than either Betelgeuse or Antares. RS Ophiuchi "went nova," hitting naked-eye brightness in 1898, 1933, 1958, 1967, 1985, and 2006. It has to be pretty close to disaster. The distance, not really known, falls somewhere between 2000 and 5000 light-years. We're unrestricted, so let's take the closer option. While farther, the Ia supernovae are (with exceedingly rare exceptions) more luminous than their core-collapse cousins, allowing RS Oph to climb to an apparent magnitude near −10, about the same as the Great Supernova of 1006.

What about double white dwarfs? Given their faintness, the critters have been elusive to say the least. The best candidate is an obscure 15th magnitude system in southeastern Sagittarius whose combined mass just hits Chandra's limit. Through some guesswork, it lies only about 400 light-years away. Were it to go, it would quintuple the wonder of an exploding Betelgeuse. But relax, the two components are probably still so far apart that their merger is millions of years off, by which time we will no longer be anywhere near them.

A Bad Tan and Other Disasters

In spite of the apparent brilliance of supernovae, we might still feel isolated. After all, even 200 light-years is a huge distance, enough to make us feel secure. At that distance the effects would be minimal. But if supernovae can go off at a few hundred light-years, why not a few tens? Cut the distance to a tenth the original, and the brightness goes up 100 times. The "death zone," within which the Earth (and its inhabitants) could suffer serious effects, is estimated—through real calculation—to be around 30 light-years, by pure coincidence, near the standard distance for absolute magnitudes. (Don't take the number too seriously. Supernovae differ considerably from one another; *the zone* could be 10 light-years, could be 50 or more depending on the kind of event. Nobody really knows.) At this distance, a typical core-collapse supernova would shine at visual magnitude −18 or so, more than 100 full-Moon's-worth of light, a white dwarf burnout at −19: imagine a sky filled with over 300 full Moons.

Given the Galaxy's supernova rate of one to three per century, the volumes of space involved, the local density of stars, the rate of high-mass star formation (which varies with time and location), and a variety of other factors, we ought to be able to calculate the odds of getting one within that critical distance. The uncertainties are such as to give a range for a 50:50 chance from as long as a few billion years to as short as a quarter-billion, the latter about the time it takes the Sun to orbit the Galaxy's center. With an age of 4.5 billion years, Earth should have witnessed at least a few such nearby supernovae and perhaps as many as 20 or even more. For much of our planet's existence, neighborly supernovae probably did not matter too much. But they surely did—and may someday again—to later complex life forms.

The effects of such supernovae are not all that direct. A nearby supernova would not turn Earth into a smoking cinder, clearly witnessed by the fact that over geologic ages life continued to progress and develop. Nor would it even strip off the Earth's tightly held atmosphere. The effects, and there is quite a list of them, are more subtle. The clearest, and first noted, is damage to the Earth's ozone layer. Ozone, an oxygen molecule made from three oxygen atoms (O_3), is created within a thick layer of upper atmosphere some 25 kilometers (15 miles) high out of ordinary breathable molecular oxygen (which has two oxygen atoms, O_2) under the action of ultraviolet sunlight. Ozone in turn fiercely absorbs ultraviolet, the strong radiation breaking it back down to O_2 and O.

We depend mightily on the fragile stuff. Sunburns and related skin damage are caused by a small bit of ultraviolet sunlight—that just to the short-wave side of visible violet—that gets through to the ground. Sunburn can be more than a nuisance; it can be lethal. Ozone effectively absorbs all the more energetic radiation, the shorter rays. Direct exposure to unfiltered sunlight in space would kill within seconds. Ozone provides us with a shield that makes life possible.

A supernova within 30 light-years would whack the ozone layer. And not once, but twice (and as seen later, thrice!). Oxides of nitrogen are death to ozone molecules. The combined X-ray/gamma-ray flash from a nearby supernova can tear apart ordinary nitrogen (N_2) molecules, producing ionized nitrogen that then combines with oxygen to make the deadly nitrogen oxides. And a little goes

Figure 8.5. The little bit of ultraviolet sunlight that sneaks through the Earth's atmospheric ozone layer is enough to burn the skin. Even the visual light is dangerous if looked at directly. A nearby supernova depleting the protective gas could allow so much ultraviolet in as to damage life in general, and could be so bright that, like direct sunlight, it too could damage the eye. J. B. Kaler.

a long way—just a small fractional increase can halve the ozone layer, which then leaks more and more especially energetic ultra-violet sunlight.

The flash would then be followed some time later by a tremendously increased flux of cosmic rays, which would have the same effect. However, while the flash is quick, the CR attack is not. The high-speed particles do not fly here directly, but diffuse outward, riding on the Galaxy's complex magnetic field, and would then hammer away at us for a century and maybe far more, doing terrific long-term harm to the ozone.

The harsh solar radiation that then gets through to the ground would produce a great increase in skin cancers, as well as other genetic damage that could markedly increase the rate of mutations. It would seriously damage photosynthesis in plants, and even penetrate through upper ocean layers to disrupt one of the great bottoms of the food chain, photoplankton, on which much of life ultimately depends. The result could easily be a mass extinction, perhaps not one so bad as produced by the asteroid impact that wiped out the dinosaurs, but bad enough. In addition, both cloud cover and lightning may be enhanced, even produced, by cosmic rays. A nearby supernova might even then create a massive climate change, adding to the effect of the ongoing "death by ultraviolet."

A close explosion may well have happened in recent geological times. One of the by-products of the nuclear maelstrom within a supernova explosion is the radioactive isotope iron-60. Still-hot Fe-60 has been found in ocean beds that can be at least roughly correlated with small extinctions that took place 3 and 13 million years ago. But remember chapter 6's "post hoc ergo propter hoc" and the Tale of the Chainsaw.

Given the age of Earth, the closest supernova that we are likely ever to have encountered would have been about 10 light-years away, enhancing all the effects by a factor of 10, giving us at maximum a night filled with thousands of full Moons.

Penguins, Soybeans, and Ants

Generally overlooked is the simple effect of a close supernova's visual brightness, which could also be a major disrupter of life. Many kinds of flora and fauna require a day-night cycle. Indeed, many are the animals that come out only in the dark. What would they make of a year or more of a no-longer-nocturnal night of 10 or 100 full Moons? Such disruptions would again lead to extinctions of many forms of life. Penguins (we can't seem to avoid them) require even more, a seasonal dark period found in the Antarctic. Aquariums that house these remarkable birds turn off the lights during the time of natural Antarctic darkness. Their lives already difficult, would these creatures survive a bright Antarctic night? Plants need day-night cycles too. A nearby supernova would have a major effect even on ordinary agriculture (as in corn and soybeans).

It gets worse. The unfiltered, undimmed Sun is too bright to look at directly without suffering eye damage. Remember frying ants on the sidewalk with a magnifying glass (or if your conscience is clear, the other kids who did)? The lens of the eye acts the same way in that it focuses sunlight on the retina, which, unless you avert your eye from the solar glare, quickly leads to retinal burns and scars.

What is not commonly recognized is that the Sun's danger to the eye is an effect of solar *surface brightness*, the amount of radiation released per unit angular *area* (per square minute of arc for example). We are told to avoid looking at solar eclipses, but not because eclipses are inherently dangerous. The phenomenon instead draws human interest, and because, even though the Sun is partially covered by the Moon, any *portion* of the solar surface is too bright to view. Even if all but a sliver of the solar disk is left uncovered by the lunar disk, that sliver can produce a burn as bad as can full sunlight. You need to cut sunlight by at least a factor of say 50,000 to make it safe to see, which can be achieved only through professional filters. (Don't even think about trying to make one.)

If you could keep the Sun at the same luminosity, but shrink it to half its present diameter, the burning effect would be four times as bad and you would need a four-times denser filter. Shrink the Sun now to a point, and it would be vastly worse. As viewed from Earth, a supernova is a point that is only smeared out by twinkling caused by Earth's atmosphere and by aberrations in the eye itself. As a result, a supernova does not have to come close to the total visual magnitude of the Sun to become visually hurtful. A simple-minded calculation that assumes that the starlight is spread on the retina over a circle one minute of arc in diameter suggests that any supernova with a brightness greater than magnitude −10 or so begins to cause discomfort, if not cause actual harm. Now try a supernova at 30 light-years, which shines at –17 or even –19. To make the effect more vivid, the surface of the full Moon appears as bright as Earth rock under full noon sunlight. Then isolate a square minute of arc, and make it 15 times brighter, as reflective as snow. Bother the eye? You bet. Such is the power of a nearby supernova that if looked at directly, even if dozens of light-years away, it could blind.

Catch a Falling Neutrino

More controversial are the neutrinos, which we explored a bit in chapter 3. The Sun sends 70 billion of them through each square centimeter of Earth (and of us) per second. And nothing happens at all.

A core-collapse supernova, however, produces around 100 trillion quadrillion quadrillion quadrillion quadrillion (1 followed by 59 zeros) of them. For a brief 10-second interval, Supernova 1987a bathed us in a flood of neutrinos the rough equivalent of those coming from the Sun. Even at their incredibly low detection rate, neutrino detectors on Earth caught about a dozen and a half: the expected number coming in from a distance of 10 billion Astronomical Units, which provided powerful proof that the core collapse really did take place.

As noted in chapter 3, there are a number of ways to catch a neutrino. One of the best is with a tub of water. The vast majority of them would just pass between and through the molecules. But on rare occasion, one will make a direct hit on an electron or proton, and kick it out at very high speed. In a vacuum, nothing can go faster than light. However, in a substance (glass, water), light travels considerably slower (the basis for ordinary refraction). If a particle exceeds the speed of light *within the substance* (which is allowed), it radiates a beam of light more or less in the direction of its motion ("Čerenkov" radiation, which gives water-bath nuclear reactors a blue glow). Neutrino hits can then be recorded by light-sensitive photocells place around the tub. The largest of such bathtubs is Japan's "Super-Kamiokande," a 50,000-ton tank lined with 11,200 phototubes.

Since life forms are made mostly of water, the same thing could happen within them. While the pickup rates from solar neutrinos (or those created by cosmic rays colliding with atmospheric atoms) are of no consequence to us, we are not so sure about the huge overload we would get from a supernova within the critical distance. Kamiokande caught about 10 of the critters from Supernova 1987a. Now bring it from 170,000 light-years to 30, and the number goes up by a factor of 30 million. An organism would now pick up about 10 neutrinos per kilogram. The resulting cell damage from the high-speed tracks could cause genetic damage, cancers, and increase mutation rates. Or not. Nobody really knows. And we are not sure we want to find out.

Shocks and Fields

A nearby supernova smacks us instantly with its electromagnetic flash (and perhaps neutrinos), then a bit later (but for a longer time) with its cosmic rays. After a few hundred years of desperately-seeking-ozone, we can relax. Comic actors get laughs with "second takes" when the joke finally dawns on them. The supernova offers an unexpected and largely unexplored "third take." And it's not funny.

The Crab Nebula, the expanding remains of the supernova of 1054, is 1000 years old. In that short time, the edge of the shock wave it has created has expanded to a radius of nearly 5 light-years, greater than the distance to Alpha Centauri. Now place a supernova at the limit to the danger zone, 30 light-years off. (Since we are only marginally sure of the details, ignore the kind, whether core-collapse or white dwarf.) Some thousands of years after the event itself, with the supernova long gone, the shock arrives at Earth. (While we get some debris from the supernova itself, hence the iron-60 in seabeds, it is so diluted by mixing with the interstellar gases that the amount is of little consequence.)

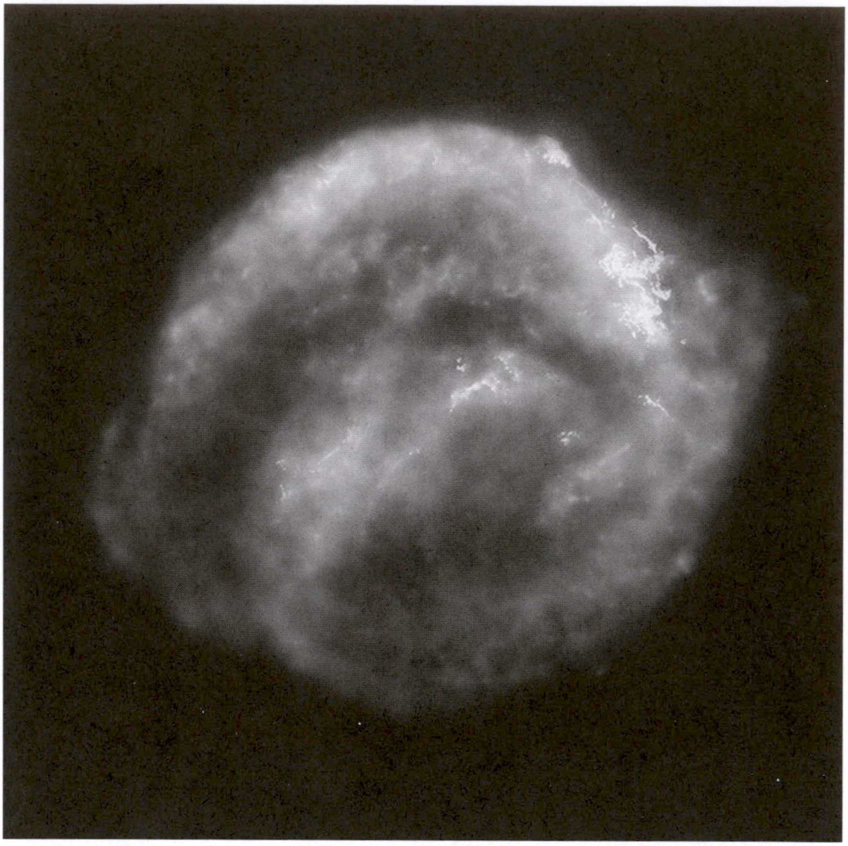

Figure 8.6. The shock wave from Kepler's supernova of 1604, seen across the spectrum in X-rays, optical, and infrared, advances toward Earth. Though now 17 light-years across, it is fortunately so far away that it poses no danger to us. Hubble image: NASA, ESA, R. Sankrit, and W. Blair (Johns Hopkins University).

The action starts when the shock encounters the solar wind. The erratic wind, enhanced by coronal mass ejections, fills our protective Van Allen belts, produces aurorae, and provides the first shield against cosmic rays invading from space. Blowing outward, the wind is finally stopped when it pounds up against interstellar matter to create a gigantic shock wave and a bubble about 200 Astronomical Units wide, giving us our heliosphere. When the supernova shock hits, it pushes the edge of the heliosphere inward on the side facing the explosion, and keeps pushing until the outward force of the wind roughly equals the inward force generated by the shock, which should happen at about one Astronomical Unit—the distance of the Earth from the Sun. The Sun, however, knows nothing about any of this and keeps pumping wind. The speeding matter now has nowhere to go but to stream out to the opposite side in a huge flaming tail. The result is a vastly distorted heliosphere, flattened on one side, flowing out to hundreds, perhaps thousands, of AU on the other.

The effects (which we can just guess at, but why not?) depend on whether the heliosphere actually compresses to inside the Earth's orbit or not, and if it does, on the direction to the supernova. If the Earth were to find itself still within a compressed heliosphere, with a vastly multiplied number of solar particles, the terrestrial magnetic field would be far more tightly compressed and the aurora would be violently visible all over the Earth. Solar particles might even make it to the ground, greatly damaging life forms. It might be bright enough at night to affect nocturnal animals.

If the compression went inside the Earth's orbit and our supernova lay perpendicular to the orbital plane, the Earth would lie entirely *outside* the compressed heliosphere. Suddenly, the Earth finds itself without its protective solar shield, and we get hit with the cosmic rays that the heliospheric bubble has kept out. Worse, we are now not just seeing the cosmic ray accelerator, we are *in* it. And there again (though no one is quite sure) goes the ozone. Unless affected by the shock or cosmic ray particles themselves, our magnetic field would also become more or less symmetric around the planet, the Van Allen belts would disappear (or be filled by cosmic ray particles), and the aurora might vanish or be significantly altered. Nobody knows what visual effects might be seen. Nobody would want to find out, either.

In between these extremes, if the supernova were to lie in or near the plane of the ecliptic, the orbiting Earth might first go into the solar wind, then out of it and into the cosmic accelerator. The magnetic field would flip wildly from nicely symmetric (when on the supernova side) to a more extreme distortion than we see now (with the supernova in back of the Sun). In any case, the results would be very visible indeed. And it could stay that way for thousands of years as the shock passes by.

And maybe more than just visible. Cosmic rays are suspected of helping seed cloud cover and of triggering lightning. What would happen were we to be "in the accelerator"? Earth could become dramatically more cloudy and stormy. Oddly, nobody knows the final effects. Clouds reflect sunlight, so we could get a "cosmic ray winter," while they also trap heat, giving us a "cosmic ray summer." In either case, even if the effects balance each other, we lose the stars. And if the Earth went in and out of the solar wind, we could wind up with two other seasons piled on top of the four that we now have.

When all the ruckus died away, we would find ourselves within a huge volume of interstellar gas heated to nearly a million Kelvin (but since the gas is so thin, like the solar corona, capable of little radiation and little harm). Indeed, we are near such a hot bubble right now, the gas most likely heated from supernovae that blew up at much greater distances in Centaurus and Scorpius, whose progenitors are related to Antares. At least if Antares were to go off, the effects on us would merely be gloriously beautiful (if you did not look at it directly), not dangerous. Any planet within 30 light-years, though, had best watch out.

As might we, given not just supernovae, but *hypernovae* that with their neutron star leavings might just be the most dangerous things of all.

Chapter 9

Hyperstars

Hyper: *prefix*, over, denoting excess;
prep, over, beyond, above; *adv*,
overmuch; *colloq*, agitated, jittery,
uncontrolled.

I t's also what you say when "super" is already in wide use and
you find something bigger and grander. In the colloquial, it's
a stand-alone word that rather well describes some of the
weirdest bodies and events ever seen. The set of definitions also
incorporates an essence of astronomy, in which we simultaneously
consider both the largest and the smallest of things, and in the
case of the stellar realm how one is transformed into the other.
Here, among the "hypers," are a collection of the strangest, largest,
smallest, most energetic, most violent, of almost anything known.
Such odd celestial beings can have such extended environments
that, even given great rarity, even given great distance, the Earth
is bound to get in the way. On the other hand, we see that Nature
may have protected us from the worst of things.

In the Navy

There has always (in historical terms at least) been a close relation between astronomy and the military. To know where one is
at sea (or in the air, even on land) long required good positions
for celestial navigation as well as accurate clock time, which is
available only by watching celestial sources, hence the U.S. Naval
Observatory and similar facilities spread around the globe. Even in

today's world, in which even automobiles navigate themselves via satellite, distant point sources of radiation (the centers of highly active galaxies that contain monster black holes) help determine the necessary precise time, telling us where the Earth is in its perpetual rotation beneath the nest of artificial orbiters.

During the Cold War, the military pumped vast amounts of money into astronomical research, the warriors and astronomers learning from each other. After all, blowing up hydrogen bombs is not a lot different than blowing up stars: to the benefit of both sides, much of the physics is the same. Los Alamos National Laboratory was—and still is—filled with astronomers, providing not just new astronomical knowledge but steady employment. Other examples abound. The Defense Department had a need to look both up (to see enemy satellites) and down (to enemy installations) with clarity, hence the development of optical systems that compensated for atmospheric turbulence. Once released from secrecy, the technology has had a powerful impact on the "de-twinkling" of starlight, computer-controlled "adaptive optics" systems now used in all major observatories.

Among the best within the .mil/.edu symbiosis was the accidental discovery of the most energetic natural events ever seen. From the mid-1960s to the mid-80s, the United States flew a suite of *Vela* (no relation to the constellation of the same name) satellites that were designed to monitor the Earth for bursts of gamma rays coming from atomic bomb explosions that would be in violation of the Partial Nuclear Test Ban Treaty. The satellites were magnificently successful, but in a completely unexpected way. Instead of seeing characteristic bursts coming from bombs (though they did find one event still shrouded in mystery), in July 1967, a *Vela* craft recorded one coming from the sky. By 1972, they had accumulated a total of 17. Nobody knew the source except that it was definitely not related to warfare.

The mystery only deepened with the 1991 launch of the *Compton Gamma Ray Observatory*, named after the American physicist Arthur Holly Compton (1892–1962). The *CGRO* was one-fourth of NASA's "Great Observatory" orbiters, the others the *Hubble Space Telescope*, the *Spitzer Infrared Observatory*, and the *Chandra* (named in part after you-know-who) *X-Ray Observatory*. During its decade-long mission, Compton's namesake found on the average of a gamma ray burst (*GRB*) a *day*. After the demise of *Chandra*, GRBs were left in the competent hands of *BeppoSAX*, an Italian-Netherlands venture launched in 1996, and *Swift*, which went live in 2004.

Arguments about origins were long and sometimes loud. Were the GRBs near or far? They were quickly found to be quite uniformly distributed across the entire sky. Since they did not stick to the Milky Way, they could not be coming from the Galaxy's disk. They *could* perhaps be coming from an extended halo surrounding the disk, but since the Sun is not centered in either, one would expect *some* kind of asymmetry. And there was none. The halo would have to be ridiculously large to smooth out the solar offset. Nor did the bursts align themselves with any prominent, relatively nearby clusters of galaxies.

The GRBs' extreme "isotropy" (same in all directions) strongly implied that instead they must be coming from huge distances,

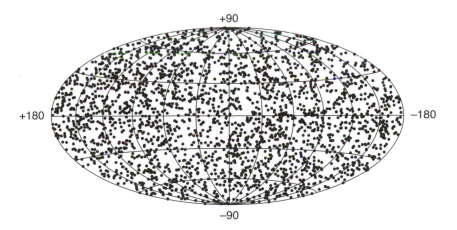

Figure 9.1. More than 2700 gamma ray bursts dot this all-sky view accumulated over a 9-year period by the *Compton Gamma Ray Observatory*. The GRBs have no preference for the plane of our Galaxy, which runs down the middle from left to right, nor for that matter, anywhere else, which tells that they are coming at us from great distances that span the visible Universe. G. Fishman et al., BATSE, CGRO, NASA.

near the edge of the visible Universe billions of light-years away. (There is no real edge, just one imposed by how far we can clearly see.) If so, however, they would have to be extraordinarily luminous to be so VERY observable here on Earth: so bright that there was no known physical mechanism for them, even the most luminous supernovae falling far—by a factor of hundreds—short.

Like nearly everything else (meteorites, supernovae), the GRBs divided into two subgroups: short bursts and long (thankfully avoiding the deadly Roman I and II). Short bursts are somewhat the more apparently energetic of the two, but are over and done with in two seconds or less, while the longer bursts last from around a couple seconds to minutes, which creates more total energy. There must be not one but two quite different explanations.

Identification

The success of modern astronomy is the result of close cooperation between space-based and ground-based observatories. From space, the stars appear still, with no twinkling. More important is that the entire electromagnetic spectrum is open to view, from infrared (blocked on Earth by atmospheric water vapor and our warming carbon dioxide), through ultraviolet (blocked by the blessed ozone), right on down to X-rays and gamma rays (also nicely cut off by the atmosphere). But satellites are so expensive, we can't make very many of them. Moreover, we can't make them really large, as we can for ground-based telescopes, which are also a lot cheaper. So from space we see with detail and across the spectrum, whereas from Earth we see in the so-called visual spectral realm (and some infrared, and of course radio), but can see vastly fainter and often with much greater speed and efficiency.

The *CGRO* could not pin down directions well enough to help astronomers stuck on the ground. But *BeppoSAX*, in conjunction with other satellites, could. In 1997, just a year after its launch, the orbiter not only found X-ray emission associated with a "long" GRB, but also nailed its location to within a couple minutes of arc. Quickly pointing a ground-based telescope in that direction,

Figure 9.2. The Hubble Space Telescope spotted the fading afterglow from the gamma ray burst of January 23, 1999, coming from the distant galaxy seen faintly above the bright burst. The whole galaxy easily fits into a tiny field of view but 5 seconds of arc (0.0014 degree) across, which gives a sense of its distance: 9.6 billion light-years. At its peak, the afterglow was 4 million times brighter than it is here, and could have been seen with large binoculars, making it the among most luminous events ever. A. Fruchter (STScI) and NASA.

astronomers saw a long-lasting (hours, even days) optical "afterglow." Such events were eventually followed even into the radio spectrum. The biggest telescopes on Earth waited. When the first known obscuring afterglows faded away, at their locations were faint galaxies. The GRBs, at least the long ones, had to be incredibly far away.

But how far? We establish the nature of the expanding Universe by correlating "redshifts" in the spectra of galaxies with their distances. In this line of work, the distances must come from objects within galaxies that we recognize and whose absolute luminosities we know from studying them in our own Milky Way system, or in other star systems that are close by. Type Ia (white dwarf collapse) supernovae, for example, fill the bill quite well, since they have similar peak absolute magnitudes of somewhat brighter than –19. Once you establish the correlation, you get the rate of expansion, now fixed well at 22 kilometers per second for every million light-years farther out.

At some point, for galaxies so far away that we can see no individual objects (you can't rely on rare Type Ia's), there is no direct way to get the distance. However, since the expansion characteristics are well established, we can just turn the problem around, and estimate the distance from the redshift: if, of course, the galaxy is bright enough to allow its light to be spread into a visible spectrum, which only the biggest ground-based telescopes can do.

Finally, GRB 970508 (that of May 8, 1997) revealed itself. Its parent galaxy had a redshift of 0.84 (the light moved to the red by a factor of 84 percent), which gives a distance of 7 billion light-years (where the distance is expressed as a "look-back time," that is, how far back in time you are seeing). Since then, large numbers of GRB galaxies have been observed to show that such distances are not only not unusual, but typical. The record is nearly 13 billion light-years. The argument was all but over: GRBs are "cosmological" in nature: they shine at us from all over the visible Universe. Even the short bursts were finally seen to have afterglows from events in distant galaxies, although in general they were not quite so far away (nicely confirming that the two kinds indeed have different origins).

Long Bursts

There remained this one small problem. The amount of energy we receive from long-burst GRBs plus their distances give the total energy release: one hundred billion trillion trillion trillion joules, or 1000 times the non-neutrino energy release of an "ordinary" core collapse supernova. There is no theory for such an outburst, no known source of such power. "Think!" said the astronomers, think, what fallacy are we holding? The answer was quick in forthcoming. You get the force of 1000 supernovae only if you calculate that the energy is being released spherically, isotropically, the same in all directions.

But nature often works quite differently. When stars are in the later formation stages and are in the process of accreting mass from surrounding disks, they consistently also *lose* mass at the same time in sets of opposing jets presumably related to their axes of rotation. Gigantic black holes at the centers of galaxies do the same thing as they consume local stars, some of their jets piercing outward for a million light-years. As giant stars lose their matter in the process of revealing their white dwarf cores, their powerful winds eventually convert themselves into similar "bi-polar flows," though nobody knows why.

So there is no reason to assume that the gamma rays, and their associated afterglows, should be spherically symmetric. If the radiation is narrowly *beamed*, coming out in twin opposing cones, we can reduce the luminosity to that of an exploding star. And sure enough, in 2003, after the afterglow faded in a distant GRB, there was the supernova. But clearly one of a different kind, as the ordinary supernovae, like the one that made the Crab, cannot be beamed, as they are observed to be nowhere near bright enough. GRBs are not just from supernovae. They are created by

Hypernovae.

The range in the birth masses of stars that produce supernovae is huge and sometimes underappreciated. They go from about 10 Suns (below which stars become white dwarfs) to perhaps up to 120 solar or even more. There is no reason to believe that over this huge mass range they should all behave the same way. We could easily expect stars of different masses to produce different kinds of supernovae and end products. And they do—we are just not very sure of what mass range does what. But we can make some good guesses.

Type II (core-collapse) SN break down into subgroups that depend mostly on how much ejecta from previous stellar winds surround the exploding stars. The brightest, which can approach the typical white dwarf (Ia) variety, are encased in rather dense shells of their own making. It's the Type I subgroups that capture particular attention. Type I's have a notoriously simple definition: they lack hydrogen. That's all there is to it. But there *are* other criteria to consider. Classic Type Ia's also have features in their spectra made by silicon. Silicon, however, is absent in Types Ib and Ic, which are respectively distinguished by having or not having helium. Ia's occur in galaxy halos where there are no massive stars, and are thus quite logically believed to be white dwarf flameouts. Ib and Ic events, however, *do* stick to galaxy planes, and therefore—like Type IIs—must *also* be core-collapsers created by massive stars. At their grandest, the Ib-Ic set can not only rival the Ia's, but can beat them, hitting absolute magnitudes of –20. Then there is the oddball brilliant core-collapser that exceeded the Ia luminosity by a factor of 10!

Supergiants are layered like onions. At a supergiant's core is the nuclear engine that may be fusing helium to carbon, or heavy stuff to even heavier stuff (ending in iron). Surrounding the core will be layers of the past dregs of fusion that include helium, all encased in a huge hydrogen envelope. The only way that Ib and Ic supernovae can be lacking in hydrogen is for the progenitor stars to have lost their outer envelopes to winds before exploding.

More than a century ago, two French astronomers, Charles Wolf and Georges Rayet, found peculiar kinds of stars that instantly became candidates for the Ib's and Ic's. Rare Wolf-Rayet stars are massive supergiants with little or no hydrogen, stars whose winds have revealed carbon- and nitrogen-rich innards that surround their nuclear-fusing cores. At the extreme, they seem even to have lost most of their helium as well: and thus the perfect fit with the Ib and Ic SN. Wolf-Rayet stars are likely born with masses above 40 or more times that of the Sun, their winds cutting them down to maybe 20 or less. As some of the original stars expanded and began to die, they became not just supergiants, but hugely luminous "hypergiants," the brightest non-exploding stars known, at maximum hitting the brightness of millions of Suns. When they do go off, they produce not just supernovae, but in some cases hugely luminous "hypernovae." All the supernovae identified with gamma ray bursts (and there are not many) seem to be core-collapse hyper-novae. Stars in the 25 to 40 solar mass range, which do not become real Wolf-Rayet stars but that still lose huge amounts of matter as red supergiants, add to the mix.

Sliding into the Pit

Type Ia supernovae annihilate themselves, so there is nothing left to view or ponder except for the expanding debris and shock wave. Not so of core collapsers, which one would think must collapse into *something*. Ordinary (how casual) supernovae, the ones from your garden-variety 10–20 solar mass stars, produce tiny neutron stars, dense balls of neutrons that pack a couple solar masses into a volume no more than 20 kilometers across, the average densities hitting 100 million tons per cubic centimeter. As the collapsing cores shrink, we once again see the conservation of angular momentum in action, resulting in fantastic spin, at least some of the neutron stars having rotation periods under a second. Powerful magnetic fields generated by the collapse and resulting rotation hit a trillion times that of Earth's field, driving immensely strong winds.

In some neutron stars, the rotation powers twin beams of radiation that then follow tilted magnetic axes. Wobbling wildly in space, if the beams cross Earth's direction, we see the neutron stars as "pulsars." The Crab Nebula is powered by one at its core, one that rotates 30 times per second and emits over the entire spectrum. Pulsars lose energy through their radiation and ultra-fast winds. In response, the driving rotation must compensate by slowing with age. After an astronomical while, all that comes out are weakened radio pulses every few seconds, and then, finally, eternal silence.

Like everything else out there, neutron stars are hardly all the same. At the extreme are very rare, slowly rotating (several seconds) neutron stars with even greater magnetic fields, up to 1000 trillion times that of Earth (1000 times that of "normal" pulsars) called *magnetars*. Different subgroups are known through their X-ray and/or gamma-ray radiation. They are probably one outcome of the massive-star collapses that creates hypernovae.

Figure 9.3. A close-up of the neutron star (pulsar) at the heart of the Crab Nebula, made by combining Chandra X-ray and Hubble visual images, reveals extraordinary energy, with twin beams emerging perpendicular to a powerful equatorial wind. Just 20 kilometers across, spinning 30 times per second, and with a magnetic field a trillion times that of Earth, the pulsar and its mighty wind power the energy of the whole nebula, which is some 9 light-years wide. As powerful as the pulsar is, it's dwarfed by the magnetars. X-ray: NASA/CXC/ASU/ J. Hester et al.; optical: NASA/HST/ASU/J. Hester et al.

Yet they are still not the small-end extreme. White dwarfs are supported by the free electrons (those stripped from atoms), which because of their wavelike behavior cannot get any closer. The vastly denser neutron stars are similarly supported by their (no surprise) neutrons. White dwarfs must be under 1.4 solar masses, the Chandrasekhar limit, else they will collapse: the ultimate cause of Type Ia supernovae, which occur when the white dwarfs are overloaded by mass gained from companions or from mergers. Neutron stars have their own limits. Since we still do not have as good a feel for the nature of the matter inside such objects (you cannot, thank goodness, take one into the lab), the limit is not as well known, though it can be constrained to between two and three Suns. Observations, which show all measured neutron stars to be below this vague limit, support the contention. If you overload a neutron star past the limit, like the white dwarf it must collapse. But there are no elements inside, no carbon to "burn," just unassailable neutrons. So all it can do is to collapse forever,

Into a Black Hole.

The subject of myths and mis-conjectures, black holes are instead some of the simplest things the Universe has to offer (creating them another matter altogether). Our planet has an escape velocity, the speed limit above which a moving body will not return, of 7.1 kilometers per second. If you could make the Earth smaller but keep its mass, everything at the surface would be closer to the Earth's attracting atoms. Your weight would increase, and so would the escape velocity. Shrink the Earth to roughly the size of a golf ball, and v-escape would hit the speed of light. Light could then not get out, and the Earth would simply disappear from sight. It would still *be* there, as would its gravity. As a black hole, it could then just not be seen.

While no power in the Universe could make our planet that small, such is not the case for stars. The Sun would have to shrink to 6 kilometers (3.7 miles) across to become a black hole. That's impossible too, as it would turn into a stable white dwarf first. But get to three or so solar masses, and you approach the sizes of neutron stars. If you could make the star a bit more massive, its gravity would squeeze it into a smaller volume. And down it goes into the pit, never to escape, never to be seen again.

In classical terms, those that would be described by Newton, the "radius" of a black hole is the *event horizon*, the spherical surface at which v-escape hits *c*. The mass of the black hole, which is still collapsing, is all inside it. Since nothing can get out, no information, we cannot describe the conditions. Classical Newtonian mechanics, however, does not really do the job. Instead it's work for Einstein, who viewed gravity as a distortion in a continuous four-dimensional spacetime, and in which a black hole is seen as an infinitely deep puncture. In a vacuum, an observer *always* sees light going at *c*, no matter the speed of the observer or of the source. Unlike a ball, you can't slow it down. But shine a beam of light upward, and as the radiation works against the Earth's gravitational field, it must still lose energy. Since it can't linger, its waves instead lengthen, that is, they redden (longer-wave light having lower energy than shorter-wave). Such "gravitational reddening" is indeed real, as it has been observed in stars. At the event horizon, light is stretched to an infinite wavelength and so loses all its energy. So the black hole goes dark.

All you need then to create a black hole is to get the core of a super/hypergiant massive enough so that, when it collapses, its mass is beyond that of the neutron star limit. Nobody knows what the birth mass of such a star must be, but an educated guess might be 40–60 solar masses and greater, into the realm of the Wolf-Rayet stars, into the realm of the hypernovae—and into the realm of the gamma ray bursts.

In the imagination, collapse the core of a really massive rapidly rotating star, one that will be well beyond the neutron star limit, one that will plunge into a black hole. As the collapse spins it up even faster, the shock wave that will drive the expansion of the hypered supernova becomes more and more concentrated toward the rotation axis. The energy of the event then emerges along the poles. Internal shocks within the expanding matter dissipate the energy into powerful bursts of opposing gamma rays that emerge within opposite narrow (2–20 degree) cones. If pointed to the Earth, we say "look at that!"; if not, we see nothing except some radiation from the dimmer supernova itself. When the shock, moving near the speed of light, hits surrounding matter (that previously lost through winds or just ambient interstellar matter), it creates the afterglow from ultrafast electrons locked into the surrounding magnetic field.

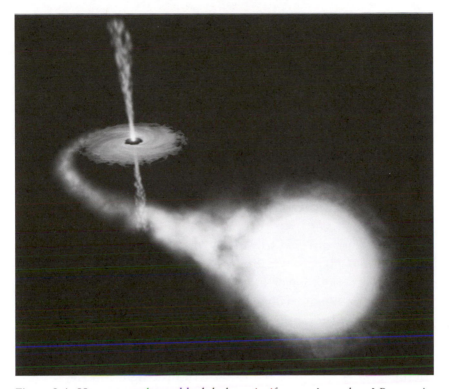

Figure 9.4. How can we know black holes exist if we can't see them? Put one in a tight double star system. The black hole raises fierce tides in the companion, which then overflows its equal-gravity tidal surface and passes mass back to the black hole. The incoming matter falls first into a circulating disk, which, because of the high compression close to the black hole, radiates X-rays, while some of the matter is ejected in twin jets. A number of observed X-ray binaries are excellent black-hole candidates. Artist's concept: ESA, NASA, and F. Mirabel (French Atomic Energy Commission and Institute for Astronomy and Space Physics/Conicet of Argentina).

The light from the supernova follows, while the core already either has become a magnetar or has plunged into the ultimate pit.

We perhaps don't even need a supernova as such. The black hole might swallow nearly everything except for the gamma-ray-producing shock, producing a "collapsar." Nothing is then seen at all but the GRB and the afterglow.

And then we come full circle. At the very extreme, remember that the most luminous supernova ever seen was actually caused by core collapse. It is suspected of being a rare "pair-instability" supernova. Even before the formation of the iron core, its internal gamma rays become so energetic that they can collide to produce paired matter-antimatter particles. The conversion sucks so much energy from the star that it collapses and, like the white dwarf Ia variety, completely destroys itself, creating not even a black hole. There seems to be no limit to the surprises and to the strangeness of the Universe.

And What Does All This Mean?

If a GRB is not directed toward us, nothing. If it is, then we may be in trouble. Hypernovae are fortunately extremely rare: else to be sure, we would never have been allowed to exist in the first place. But we might easily expect to get hit at least once, maybe more. The most important criterion is the "danger zone," which for ordinary supernovae is (within a large uncertainty) about 30 light-years. The major factors are a huge gamma ray burst that in total energy is far larger than that from the light (taken over the spectrum) of the burst from your common supernova, which puts the death zone at a far greater distance than before. The major consideration, though, is the beaming. All the energy is concentrated into a narrow cone, the width of which varies considerably from one to the next, but on the average would increase the critical distance to around 6000 light-years!

   Six thousand light-years defines a monstrously large volume that contains a large number of massive stars, including some of the most massive known. The saving grace is GRB rarity. An ordinary SN should pop off in its small 30-light-year danger zone at best once every 250 million years. (Remember, this number is very uncertain. We could also get two in a row and another not for more than a billion years.) Given the rarity of the highest mass stars, we might expect one in any large galaxy like ours every 100,000 to a million years, as opposed to one or two a century for the ordinary version. But now the beaming works *for* us rather than against. For a GRB to be dangerous, the blast must be pointed almost directly at us, which reduces the chance of getting hit by a factor of several hundred. The best estimate is that we might get one of them within the danger zone every billion years or so. Again it's a 50:50 chance. Statistically there are small odds for many more, decent odds that we have *never* gotten it. And the odds of not getting hit by closer ones get better very fast with decreasing distance.

The effects would be similar to those of a close ordinary SN. The main one would once again be to strip the ozone layer, causing harsh ultraviolet sunlight to stream down through the atmosphere, in turn causing a mass extinction. Cosmic rays generated by the hypernova shock would probably not be much of a factor, however. We are saved from them by the vast distances involved and their subsequent diffusion by the Galaxy's magnetic field. Some astronomers think that ultra-high-energy cosmic rays can be produced by GRB/hypernovae, but there are so few of these fleeting energetic ions that they pose no danger to us.

The billion-year even-odds time frame encompasses all the big five extinction events. The K-T event of 65 million years ago (and the death of the dinosaurs) is pretty clearly the result of an asteroid impact. The Permian event of 245 million years in the past, that of the Great Dying, may have been the result of huge volcanic action that created the Siberian volcanic floods, though the newly found Antarctic impact may be a suspect. The first of the Big Five, though, the Ordovician of 440 million years ago, seems to some experts to be the best possibility of having fallen to a GRB, though we may never know.

And in the Future?

Earth has another 5 billion or so years in front of it before the evolution of the Sun to a giant state pretty much does it in. Life on Earth probably has a shorter span, however. As the fuel supply in the solar core diminishes, the Sun is ever-so-slowly brightening. We thus have perhaps only a couple billion years left before the oceans are cooked away. (We should not be depressed. While the federal budget throws "billions" around like gum wrappers, making the

number seem trivial, it is not. You would not want to count to a
million, and a billion is a thousand times that!) Two billion years
is plenty of time within which we could "catch a failing star." Or
two or three, or none at all.

What about the near future? What's around us that could be dangerous? The champion Galactic hypergiant is drably named Eta Carinae in the southern constellation Carina, the Keel of a much larger
older constellation, Argo, the Ship of the Argonauts. Eta is tucked
into a massive interstellar cloud so bright and obvious that it is visible to the naked eye. Around 1846, the star blossomed in an extraordinary non-explosive eruption that made it the second brightest
in the sky, only Sirius beating it out. By the turn of the century it
had faded back to under naked-eye visibility as it surrounded itself
with a huge, dusty cloud of its own making. When studied across
the spectrum, this quite-hot hypergiant may be the most luminous
star in the Galaxy, radiating at a rate of some 5 million Suns. And
it's not alone. Subtle variations in the spectrum and in its X-ray
radiation strongly suggest a lower mass companion that orbits in
5.5 years. Eta itself, the visible star, was probably born with a mass
well in excess of 100 Suns. Losing vast amounts of matter, tens of
Suns, it turned itself into a hypergiant and may now be becoming a
massive Wolf-Rayet star before blowing up as a hypernova. Perhaps
it will even skip the W-R step. We don't really know.

At a distance estimated between 6700 and 8000 light-years, Eta
Car is near the rough edge of being in the critical volume. It could
literally go off anytime, and even though slightly beyond the above
danger zone (which is very uncertain anyway), it could still do
harm. It may produce a black hole and a gamma ray burst. But
once again, the saving grace: one of the twin beams would have to
be pointed at us. And from what we know about Eta Car (which
is actually quite a bit), the rotation and orbital axes, which is the
most likely direction in which the blast would extend outward, are
clearly pointed toward somebody else. Poor them. On the other
hand, it might make a grand pair-instability version. All we can do
is speculate.

Eta Car is not, however, the only one around. Given its promi-
nence, set the danger zone at its upper distance limit of 8000 light-

Figure 9.5. More than a century and a half ago, Eta Carinae, a double star that contains one of the most massive and luminous stars known, ejected more than six Sun's-worth of itself, which is now seen as a huge expanding bi-polar dusty cloud. The star could explode, perhaps as a hypernova, almost any time, creating a magnetar, a black hole, or in the extreme, nothing at all. Hubble Space Telescope, Jon Morse (University of Colorado), and NASA.

years. Here are nine more stars with birth masses above 40 solar within that rather arbitrary limit. Marching from the edge of the danger zone toward us are the following:

- fifth magnitude Rho Cassiopeiae (40 Suns, 8000 light-years)
- fifth magnitude P Cygni (50–60 solar, 6000–7000 light-years)
- fifth magnitude Zeta-1 Scorpii (60 solar, 6000 light-years)
- 12th magnitude Cygnus OB2#12 (over 60 Suns, 6000 light-years)
- eighth magnitude VY Canis Majoris (maybe 40 Suns, 5000 light-years)

- fourth magnitude Mu Cephei (Herschel's Garnet Star, 40–50 Suns, 2400 light-years)
- fifth magnitude VV Cephei (40 solar, 2400 light-years)
- second magnitude Zeta Puppis (60 solar, 1400 light-years)
- second magnitude Gamma Velorum (double with a massive dwarf and a Wolf-Rayet star, 1200 light-years)

There are several more between 10,000 and 20,000 light-years as well, and these do not include a large number of massive stars near the center of the Galaxy that remain largely unexplored. Any of them within 8000 light-years could affect Earth. Note especially the distances of the last two on the list. Of course, again the beam would have to be pointed toward us, which makes the chance that one of the eight will get us to (rounding off a lot) 1 in 100.

Some of these massive stars are pretty unstable: witness Eta Car. In 1945, Rho Cas belched a dusty cloud that changed its color for several years. Four hundred years ago, P Cygni, which loses mass at a rate 10 billion times that of the Sun, ejected mass in the mode of Eta Car. Brightening by a factor of two or three, it also carries the name "Nova 1600 Cygni." While at best at the 40 solar mass limit, the red hypergiant VY Canis Majoris radiates at the rate of half a million Suns, is erratically losing mass, and with a radius near that of the orbit of Saturn, is one of the largest stars known. Both VV and Mu Cephei are erratic variables, and VV (near VY Canis Majoris's size) is affected by a close massive companion. Cyg OB2#12 (a catalogue name) is one of the most luminous stars of the Galaxy. If not for huge dimming by intervening interstellar dust, it would be first magnitude and have a name. While interstellar dust would dim the optical afterglow, the gamma ray bursts (which don't "see" the dust) would get through just fine. Gamma Velorum contains a W-R star. No more need be said.

Afterglow

The word sounds so calm and benign. So peaceful. Hardly. Rather like the brilliance of a nearby ordinary supernova, the optical afterglow is an underappreciated danger. While the afterglows show great differences among themselves, they can be amazingly bright. The famed gamma ray burst of January 23, 1999, hit ninth magnitude: from a look-back distance of 9 1/2 billion light-years. It would have been visible in a small backyard telescope. Put the thing much closer and it would have been visible to the naked eye. Another seen in red light in 2006 was twice as bright. The champion, 2008's GRB 080319B, hit between magnitudes 5 and 6, and indeed *was* a naked-eye object. It was 7.5 billion light-years away. Various transient brightenings might be similarly so explained.

While most afterglows do not reach these levels, as long as we are speculating (a bit wildly, but why not), take the above list of candidates and see how bright the afterglows might be were they to create a particularly bright one. Try Eta Car. Move GRB 080318B to 8000 light-years. For simplicity, ingore any interstellar dimming and the fact that the light of the glow has been shifted into the optical from the ultraviolet compliments of the expanding Universe. The results are so extreme that even big corrections are irrelevant. At Eta Car's distance, the glow would appear roughly as bright as the Sun. Such a blast from a point source would be instantly blinding. At the distance of our nearest hypernova candidate, the maximum afterglow would well exceed the apparent solar brilliance. Night would be brighter than day, night *would* be day, all from one destroyed star just over 1000 light-years away.

And as long as we are speculating, what about one within the "ordinary" death zone of 30 light-years? Given the low numbers of candidate high-mass stars, the odds are extremely low. But never zero. The maximum afterglow from such an explosion would (give or take, who would care?) hit apparent magnitude −40, 150,000 times the solar brightness. The Earth, though not destroyed, would be reduced to a blistered cinder, most, if not all, life forever gone. We are blessed that nature does not much like to make super-high-mass stars. Or we would might not be here. Neither, for that matter, would anybody else. Anywhere.

But perhaps we can relax, as Nature herself may provide protection from these most devastating celestial events.

Safety Zone

Not enough is yet known about the environments of hypernova gamma ray bursts. The distances of most are so immense that we do not have firm statistics on the parent galaxies. As stars form and age, they manufacture heavier chemical elements from lighter ones and, through their winds and explosions, send them back to interstellar space, where they find their ways into the next stellar generation. The heavy element content of a galaxy thus increases with time. There seems to be a notable tendency for the bursts to appear in small, irregular galaxies in which the various processes of stellar evolution have not been able to boost the metal content very high. The sheer size and age of our own system then, wherein supernovae have been increasing the iron abundance for billions of years, might itself protect us from serious harm.

But what about "others"? We may not be alone in the Universe. For all we know, and it is precious little, there may be other stars with planets teeming with life. Someone, somewhere, may indeed witness one of these amazing detonations up close, so we might out of sympathy at least try to appreciate the consequences. Yet "they" seem to find some protection too. There is strong evidence that planets tend to form around stars that are, like the Sun, metal-rich. Life and GRBs then may be mutually exclusive. If nothing else, the above exercise in unleashed destruction reveals just how powerful Nature can be.

But even if we are protected here from long-burst hypernovae, there are always the short bursts.

Quick Time

The destructive power described so far is only for the long bursts. What about the short GRBs? They are not just lesser versions of the long bursts, but instead come from a different population altogether, with almost certainly a different cause. Though at cosmological distances, their parent galaxies (when identified by modest afterglows) are generally closer, and it's not clear that the bursts are all beamed like those of the longer GRBs. Though their gamma rays are on the average "harder," there are fewer of them since the bursts are so transient. No supernovae are seen either.

Since there is no "smoking gun," we can only speculate. One scenario for the explosion of a Type Ia supernova is the merger of the members of a binary white dwarf. Why not neutron stars? Start with two massive stars in a binary system. As the more massive evolves into a supergiant, it enfolds its partner with its mass-losing wind, or even its outer layers, which draws them together. It then proceeds to blow up as a core-collapse Type II supernova, turning into a neutron star. Under some circumstances (off-center blowup for example), the lesser of the pair may get kicked out of the system. A number of high-velocity "runaway stars" are ascribed to just such a fate. But a fraction (which we know not) will survive as a couple. Then the second one evolves, envelops the neutron star, bringing them yet closer, and then itself blows up. Now we have two neutron stars in a close binary.

As the two orbit, gravity waves sent into the fabric of space-time cause them to lose orbital energy. Spiraling ever closer, they finally merge—and yet another short GRB is sent rocketing across the cosmos. Neutron star–black hole mergers might work as well, while white dwarf–neutron star mergers are suspected for some long bursts. Binary pulsars, neutron stars that carry their own internal precise timing engines, are indeed observed, and are seen to change their orbits quite in accord with prediction (and as an aside, supporting gravity-wave theory and relativity in general). So it can happen. While not as dangerous as the long GRBs, short bursts are still plenty powerful, and you would not want one to go off anywhere near you. Fortunately, the closest candidate seems to be tens of thousands of light-years away. Duplicity, however, may not be necessary.

Magnetar: The Future Is Here

While supernovae, hypernovae, gamma ray bursts, and the like all have potentially disastrous effects, so far as we know, they pose no immediate threat: we've considered only what may have happened in the distant past or might take place in the far-off future. While less well known and overall less powerful, magnetars have the great draw of being of the here and now.

We know of well over 1000 pulsars. The number of quiet neutron stars must be vastly more. Even at the present rate of star formation, given the 10-billion-year life of the Galaxy's disk, there should be at least 100 million of them. There is probably one nearby, sliding silently past us, of no danger whatever. The tiniest fraction, fewer than 20 known, are the extraordinary magnetars, which have magnetic fields so powerful as to be lethal at a distance of 1000 kilometers. Though they ally themselves with the remnants of supernovae, how they are created is uncertain. One strong suggestion is that they are the progeny of the most massive of stars, which lose so much matter through winds as they evolve as hypergiants that they don't have enough mass left to become black holes. If the progenitor stars were rapidly spinning so as to create the intense magnetic fields, they turn into magnetars instead. But, once again, we don't really know.

We *do* know the effects they can have, however. As the spins of ordinary pulsars slow (the one in the Crab being a prime example), they undergo periodic "glitches" in which the rotations suddenly—and temporarily—speed up. The cause is thought to be a relatively modest "starquake," in which the strong magnetic fields re-adjust the neutron-star outer crusts. Magnetars, with magnetic fields up to 1000 times those of ordinary pulsars, take this behavior to an extraordinary extreme.

As rare as magnetars are, they have an even rarer subset known as "soft gamma-ray repeaters," or (to add to the alphabet soup) SGRs. Only five are known, and one of these is in our nearby companion galaxy, the Large Magellanic Cloud. Brief X-ray pulses reveal them to be long-period (many seconds) pulsars, placing them in the magnetar clan. If the Crab pulsar has a thing called a "starquake," it's hard to know what word to use for what these things do.

On August 27, 1998, SGR 1900+14 in Aquila (the numbers are coordinates) launched one of the greatest stellar attacks ever seen

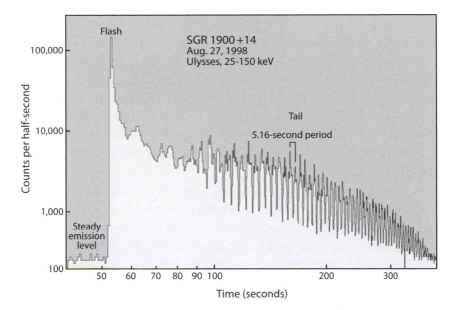

Figure 9.6. A "superburst" from the "soft gamma ray repeater" SGR 1900+14 powerfully affected the Earth's atmosphere even though it was 20,000 light-years away. For a brief, shining instant, it radiated considerably more power than our entire Galaxy of 200 billion stars. As the burst faded, short variations revealed the rotating magnetar within. Kevin Hurley, University of California at Berkeley, and NASA/Marshall.

on Earth, from a mega-starquake, in which twisted magnetic fields attempting to re-align themselves cracked the magnetar's crust. In a pulse that lasted less than a second, the resulting flood of gamma rays hitting us overloaded satellites' electronics, and most amazingly ionized the Earth's upper atmosphere. Within a few hundred seconds, the event and its aftermath were all over with. SGR 1900+14 is 20,000 light-years away.

Reprising Jolson's famous phrase from the last chapter, on December 27, 2004, SGR 1806-20 (in Sagittarius) outdid its Aquila cousin by a factor of 100. For a brief instant, a couple tenths of a second, an outburst of energy the equivalent of half a million years'-worth of sunlight shone on us with the apparent light of more than a full Moon (in gamma rays: you could not have seen it). Once again, many satellites and the upper air took a huge hit.

SGR 1806-20 is estimated to be *50,000* light-years away, on the other side of the Galaxy. And it was not the first time! The magnetar had previously popped in 1974. And it will most likely do it again before these pages turn brittle with age. Magnetar bursts are seen in other nearby galaxies, and may constitute 10 to 20 percent of all short GRBs. Yet because the blast is so brief, the death zone is a mere few light-years. While the effects are noticeable, the magnetars are too rare to be a real danger. Except of course maybe in the distant future.

Rain, snow, storms, bring disaster to us. They are a focus of daily life. Yet without the storms, we would have no life at all. Stars and supernovae are no different. While they have the potential to bring catastrophe, they too bring life. We would not be here without them. The balance between creation and destruction seems to be just right. Read on.

Chapter 10

Coming Home

Our journey began by reaching out to the cosmos to see what is there. We now return home, to Earth, as we witness the treasures that the heavens have brought us. We no longer look at direct, specific influences, though there are many, and undoubtedly many others that have not been brought up or even yet recognized. (We enter a dark interstellar cloud and the Sun dims enough to chill Earth; we inadvertently get in the way of a gigantic laser beam fired by space aliens . . .) Leave the disasters behind. Instead, focus on the pervasive undercurrent of these chapters, that we are the product of billions of years of stellar life, indeed of the whole Universe. It's time now to savor the overall gift of the sky, of the heavens, of the cosmos, to examine the raw material out of which we are all made, the raw material of life itself and how it came to be.

In the Beginning

There was the Big Bang. Just short of 14 billion years ago, the matter—the energy really—of the Universe was compacted into a hot dense state of no known dimension. Where it came from nobody knows. Since there is no information that could survive from that most primitive time, this ultimate mystery may never be revealed to us. What *is* revealed is the spectacular aftermath.

For reasons that we do not understand, the hot condensed soup of a just-created Universe took off in a sudden, extraordinarily fast expansion, and like any expanding medium, immediately began to cool. Energy and mass are two aspects of the same thing. We can convert mass into radiant energy—as happens within stars. Conversely, if the energies of radiation are high enough, the process can be reversed. While the Universe cooled, violently colliding gamma rays turned part of their energy into matter (and antimatter, but matter won), into the protons (hydrogen nuclei), neutrons, and electrons of today, all awash within a bath of continuing high-energy gamma radiation.

In the expansion, the leftover gamma-ray waves, caught in the web of spacetime, stretched out. Losing so much energy that they could no longer create mass, they paraded through the spectrum, eventually turning into X-rays, ultraviolet, then visual, infrared, and finally—in our own era—into radio waves. They are all around us today as the *Cosmic Microwave Background*, the radiation providing a faint electromagnetic hiss against which we broadcast our artificial signals, radiation that has come to us from the Universe nearly in its ultimate primitive state. Big Bang theory predicts a temperature to the Universe of 3 degrees Celsius above absolute zero. And that is just what we find (actually, an amazingly precise 2.725 . . . degrees), offering us some of the best proof available that the Big Bang really did take place and pretty much voiding any competitive notions.

While high temperatures in stars promote fusion of light elements into heavier ones, within the first three minutes of existence (a word fraught with danger, as there may well have been some kind of "existence" before our very own Big Bang), the temperature was *too* high. Any attempt at fusion would have been foiled by destructive atomic collisions. But then for a brief, shining moment, the expanding, cooling Universe passed through the temperature of a stellar interior, first from billions of degrees down through millions. Protons could grab neutrons to combine permanently into deuterium (hydrogen with a neutron attached). Then with the deuterium, further reactions created the isotopes of helium (helium-3 with two protons, one neutron, helium-4 with two of each), then lithium-7 (three protons plus four neutrons). However, progress to iron, as happens in a massive star, takes ever-*in*creasing temperatures, not ever-*de*creasing ones. Rapidly declining, the Universe's temperature simply dropped too low to make anything else, leaving us stuck with only the lightest of chemicals.

Figure 10.1. An all-sky map of the near-perfect Cosmic Background Radiation (the sky's equator stretching from right to left) glows at 2.7 degrees above absolute zero. Released half a million years after the Big Bang when the expanding mix of mass and energy first became transparent, it is the oldest thing we know. Hundred-thousandths of a degree variations, observed here by the Wilkinson Microwave Anisotropy Explorer Satellite, make the subtle ripples that became the strings and clusters of the galaxies of today. NASA/WMAP Science Team.

Stars have made all the rest. When we look into the most primitive bodies we can find and allow for some effects of stellar aging, cosmologists predict that for every million hydrogen atoms, the Big Bang should have made 83,000 atoms of helium, 28 of deuterium, 9 of helium-3, and 0.0004 of lithium. And (with the exception of lithium, which is a difficult case) again that is what we find, yet more powerful evidence for that greatest of all creative events.

Ancient Stars and Forming Galaxies

In the radio spectrum, the whole sky is alight with the Cosmic Microwave Background. Look closely enough and we see it break into subtle "ripples" whose character is fully supportive of the Universe's expansion rate, age, quantities of dark matter and dark energy (all spelled out in chapter 1), and primitive chemical composition. They

Figure 10.2. At left, 70,000 galaxies observed in the Sloan Digital Sky Survey (the Sun at center) march off toward the distant visible horizon of the Universe. Distances derived from redshifts out to 2.5 billion light-years, in wedges centered on the sky's equator, reveal (at right) filaments, walls, and clusters that relate to variations in the Cosmic Microwave Background. Astrophysical Research Consortium (ARC) and the Sloan Digital Sky Survey (SDSS) Collaboration, http://www.sdss.org.

also reveal the first glimmerings of the formation of lumps in the distribution of matter that will eventually turn into the clusters of galaxies, then the galaxies themselves, of today.

How and when the first stars formed is unknown. Perhaps they developed within primitive galaxies that were growing through mergers of the lumps in the early Universe. Perhaps they condensed on their own in what was then open space, acting as the cores that drew in surrounding matter to begin the construction of new galaxies. However they were created, they must have been made only of the three lightest elements, hydrogen, helium, and lithium (the last so rare we can, for now, ignore it).

In our own time, we look at the chemistries of stars through analysis of the "bar codes" in stellar spectra. Lower mass stars have such long life expectancies, longer than the age of the Universe, that we expected to find some of the original stars, those

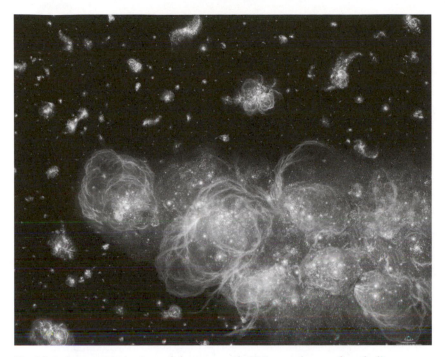

Fig 10.3. In an artist's view of the very early Universe, hosts of exploding massive stars seed the cosmos with the first heavy elements. Painting by Adolf Schaller for STScI; science background by NASA and K. Lanzetta (SUNY).

made from just the lightest elements, the "zero-metal" stars. We can't. We can go back in time to when stars had *almost* no metals, with iron-to-hydrogen contents under 1/100,000 that of the Sun. But we can't get to zero. Is the theory wrong? Did the Big Bang not really happen? Which is a worst-case scenario, as then we'd have to think of something else, and there *isn't* anything else.

Or are there simply no zero-metal stars to be found? Star formation today requires dust to cool interstellar molecular clouds. In the absence of dust, developing stars require a different cooling mechanism, one from the hydrogen alone (in its molecular form), which from theory seems to promote *massive* stars, 100 solar masses, 200, even more, that would have blown up long ago, leaving us with no stellar remnants from that archaen age other than the massive black holes into which they collapsed: black holes that grew by feeding off surrounding matter and that may have become the supermassive black holes we now see residing at the cores of large galaxies.

As the primitive stars evolved, their massive nuclear-burning cores would have turned first to helium, then carbon and oxygen, following much the same nuclear path as do the massive stars of today. Winds furiously blew new elements into the cosmos. Core collapse or pair-production exploded them. And in the nuclear fury, they burned themselves all the way through the periodic table, through all the elements, salting the interstellar hydrogen with the first really heavy atoms. Dust could now form through gas condensation, and stars could now begin to be born in a more normal manner. We can't find the zero-metal stars because there aren't any left, or in the case of lower masses, were never there to start with. Here is where we begin our last journey among the stars, one that will reveal how the stuff we are made of was born.

At the Table

If chemists have an icon, it's probably the periodic table that so many came to know and love (or loathe) in high school. It's a brilliant organization of the chemical elements, in which the number of protons increases steadily along the rows, while the columns have similar chemical—molecule-building—behaviors. While chemists and others take the elements nature gave to us and learn how to join them into new substances, astronomers are more interested in working backwards to see specifically where they all came from, to learn of their ultimate origins.

Starting with the Big Bang's hydrogen as the raw material, astronomers then strive to replicate one of the great fundamental data sets of science, the chemical composition of the Sun, hence Earth. Ignoring the cores of dead and dying stars, by number of atoms, hydrogen completely dominates at about the 90 percent level. Almost all the rest is helium. Subtracting these two leaves a tiny residue, a mere 0.15 percent, which contains all the other elements. Unimportant? Hardly. Earth is made from this remainder.

1 H																	2 He
3 Li	4 Be											5 B	6 C	7 N	8 O	9 F	10 Ne
11 Na	12 Mg											13 Al	14 Si	15 P	16 S	17 Cl	18 Ar
19 K	20 Ca	21 Sc	22 Ti	23 V	24 Cr	25 Mn	26 Fe	27 Co	28 Ni	29 Cu	30 Zn	31 Ga	32 Ge	33 As	34 Se	35 Br	36 Kr
37 Rb	38 Sr	39 Y	40 Zr	41 Nb	42 Mo	43 Tc	44 Ru	45 Rh	46 Pd	47 Ag	48 Cd	49 In	50 Sn	51 Sb	52 Te	53 I	54 Xe
55 Cs	56 Ba	57 La	72 Hf	73 Ta	74 W	75 Re	76 Os	77 Ir	78 Pt	79 Au	80 Hg	81 Tl	82 Pb	83 Bi	84 Po	85 At	86 Rn
87 Fr	88 Ra	89 Ac	104 Rf	105 Ha													

58 Ce	59 Pr	60 Nd	61 Pm	62 Sm	63 Eu	64 Gd	65 Tb	66 Dy	67 Ho	68 Er	69 Tm	70 Yb	71 Lu
90 Th	91 Pa	92 U	93 Np	94 Pu	95 Am	96 Cm	97 Bk	98 Cf	99 Es	100 Fm	101 Md	102 No	103 Lr

Fig 10.4. The Periodic Table organizes the chemical elements horizontally in order of proton number and vertically according to their chemical properties. The filled boxes are those elements observed in the Sun via their spectral signatures or directly in the solar wind. Some elements are so radioactive that they effectively don't exist. Others are in such low abundance (like uranium, no. 92) that they escape detection. J. B. Kaler, *Cambridge Encyclopedia of Stars*, New York: Cambridge University Press, 2006, reprinted with the permission of Cambridge University Press.

Put the Sun into a big still, boil off the hydrogen and helium, and the leavings are what the Earth and other small bodies are made of, including asteroids and meteorites. As we go up in proton number through the periodic table, the abundances of the individual elements decline (with even numbered being more common than their odd-numbered neighbors). Then we see a rise in numbers around iron, element 26, followed by a steep drop and another general decline. Using the theories of stellar and Galactic evolution as well as those of nuclear physics, out comes a good approximation to the solar composition. It all works reasonably well, suggesting that astronomers really do know what they are doing.

In the broad picture, all the chemical elements except hydrogen are manufactured within stars or made by stellar action on the gases of interstellar space. Stars manufacture elements by a variety of processes. In dying stars over much of the whole mass range, convection brings the newly forged atoms into the upper stellar envelopes, which are then lost through winds. Like the first stars, high-mass stars still make elements explosively through supernovae and violent ejection, as do exploding white dwarfs in double-star systems. The newly made elements then pepper the gases out of which later generations of stars will be born, the heavy element content of the whole Galaxy increasing with time.

Hydrogen comes to us from the Big Bang. Helium has a special place. While the bulk of our helium was made in the Big Bang, vast amounts of it are still created by hydrogen fusion in the cores of stars. Most of that is locked up forever or converted into carbon, so we can ignore it. But in evolving giant and supergiant stars, some of it is indeed brought upward to be released. These and the ejecta from supernovae have added about 25 percent to the total free helium content of the Universe, bringing the number up to about 10 percent of that of hydrogen. (A tiny exception: Earth's helium is made by the decay of uranium atoms. But since the uranium was made in supernovae in the first place, the rule that the combination of the Big Bang and stars make it all is preserved.)

Lithium, the other Big Bang element, is also a curious case. It's been altered as well. While not made *in* stars, it is created from cosmic rays (which are made *by* stars) smashing into interstellar carbon and oxygen, as are beryllium and boron (4 and 5 protons). These elements, though, are fragile, and are easily destroyed through nuclear processes in the outer envelopes of stars (the Sun's lithium nearly nil), making it quite difficult to check the amount that the Big Bang offered us.

The Sun's internal hydrogen fusion involves the direct collisions of protons that build to helium. More massive stars, those above about double solar, do it differently. At their higher internal temperatures, it's more efficient to use carbon as a catalyst. Successive collisions with protons and quick radioactive decays create various isotopes of carbon, nitrogen, and oxygen before finally spitting out a helium nucleus, with the carbon returning back to normal (the process called the *carbon cycle*). Much of our world's nitrogen was made in this way.

Of course, the carbon had to come from someplace. It's produced in abundance by the fusion of three helium atoms within advanced giants before they eject their outer envelopes into space. Some of it gets into the ejecta, and provides the eventual raw material for the nitrogen-producing carbon cycle. The first carbon was made by the zero-metal stars that are no longer with us. Most, or at least much, of the carbon of which life is made is the result of later generations. Capture of helium nuclei, with their paired protons, explains the up and down pattern of solar abundances, in which the evens trump the odds.

Proof? In the Sun and elsewhere, oxygen is about twice as abundant as carbon, true in the Sun and in the rest of the Solar System. Dotted around the sky, however, are deep red stars in which the ratio of carbon to oxygen is completely reversed and then some (carbon molecules strongly absorbing blue light, rendering the stars blood red). More massive stars eject nitrogen and carbon, as well as other elements, through the winds that turn them into the stripped Wolf-Rayet stars, allowing us in a sense to see the nuclear action in progress.

These and other massive stars then explode as supernovae, even hypernovae. Within the shocked expanding envelope surrounding the forming neutron star, in a cauldron heated to more than 10 billion degrees, uncontrolled nuclear fusion madly begins to create all the elements. The "great burn" is trying to advance to the most stable elements of all, iron and nickel. If the supernova could just sit there long enough, that's all we would get. But the star is also violently expanding and cooling, which squelches the burn before it all goes to the two critical elements. The result is a truckload of oxygen, silicon, and a variety of other things launched into space. Nearly all the oxygen you breathe was made in this way. Core-collapse supernovae each produce about a tenth of a solar mass of pure iron. Type Ia's, the white dwarf variety, do the same thing, except in an even grander way, each creating about three times as much iron.

The remarkable beauty of it all is that we can actually see it happening. The visual brilliance of white dwarf supernovae is the result of the decay of radioactive nickel into radioactive cobalt and then into stable iron. The fading rate of such a supernova is quite in accord with the decay of these (and other) unstable isotopes as they heat the gas. After the initial blast, Type II core collapsers reveal the same thing. Even the resulting gamma rays have been detected. The outcome is that, except for magnesium, iron is the most abundant of metals. It's *all* been made in supernovae.

Iron has but 26 protons. Uranium has 92. To explore this huge gap, with all its important, though relatively rare, elements, we need those wonderful, free (no charge)

Neutrons.

Whatever the process, nuclear fusion in stars takes heat, a lot of it, to get the nuclei moving fast enough to overcome the repulsive force between the positive protons in each of the colliding nuclei. The resistance of atoms to fusion is a good thing, really, as otherwise stars would simply all blow up as gigantic hydrogen bombs. Neutrons, from their very name, are electrically neutral. They feel no repulsive force, and can thus ram right into an atomic nucleus and change its character. Like supernovae and gamma ray bursts, there are two kinds of neutron capture: fast and slow. It's not the speeds of the neutrons that count, but the rates of capture. And the two sorts are terribly different, with no overlap.

Slow capture (the "s-process") is the better understood. An advanced giant star has a dead carbon-oxygen core that is surrounded by a shell that fuses helium to more carbon, which in turn is surrounded by one that fuses hydrogen into helium. The helium-fusing shell, where free neutrons are nicely liberated in other reactions, is the source of the action. Best to take an example, say the most common isotope of zirconium, Zr-90 (40 protons, 50 neutrons). Wandering through the hot dense gas, a Zr-90 atom is struck and penetrated by a neutron, which simply attaches itself via the powerfully attractive strong force, turning the zirconium into Zr-91.

Which is also stable, so it can wait a long time before it gets whacked by another neutron, making it Zr-92. It's stable too, so eventually, in the heat of the action, it gets hammered into Zr-93.

But that's too many neutrons for internal relations in the nucleus to remain in a happy state. To squelch the family feud, one of the neutrons ejects an electron. The donation to the atom's surroundings turns the philanthropic neutron into a proton, and thus the nucleus into an isotope of the same weight, but with the next higher proton number. Zr-93 thus becomes niobium-93 (41 protons). It's also possible that before the neutron shoved the electron out the door, the unstable Zr-93 might take another neutron into the house to make again-placid Zr-94. And so on, until we hit Zr-98, in which the family becomes *so* contentious that it *all* goes to niobium-98.

Now just continue the process. Nb-93 collects another neutron, making Nb-94: which decays through electron ejection into molybdenum-94 (42 protons), or gets another neutron to make Nb-95. Etc. In this way the niobium is turned into molybdenum, Mo into technetium (43 protons), Tc into ruthenium, etc., all the way to bismuth (83 protons). There, the process comes to a halt, as heavier nuclei are all radioactively unstable and can't live long enough to pick up the invading neutrons.

Technetium is special, both in the world of the nucleus and in the hearts of astronomers. Because there isn't any. At least on Earth. All of its isotopes are radioactive, the most stable lasting only a few hundred thousand years. If there had been any when the Earth was born, it's surely all gone now. (There is always an exception. Tc, valued for its very instability, is made in Earth labs for medical purposes; injected into the body, it's great at detecting cancers and thyroid problems.)

But there it is in the outer layers of a selection of giant stars. The *only* way it could be present is to be made within the star and brought to the surface. Further proof of the process is that elements made by slow neutron capture are in the calculated ratios in the Sun and in meteorites. From the abstract to the particular, the s-process makes not just nearly all your zirconium (valued in alloys), but most of your strontium (fireworks), lead (weights and sinkers), tungsten (bulb filaments), and a huge variety of other materials needed for daily life.

But that leaves us hanging at element 83. For the rest we call on

Rapid Capture.

Or the "r-process." Rapid neutron capture itself is pretty well understood. Slow neutron capture ends when an isotope of any particular atom becomes so unstable that it cannot wait long enough to take on another neutron boarder. But if you hit an atom with a neutron machine gun, one neutron pounding in after the other, you can build *very* high-order, and *very* unstable, isotopes. Until you get to one that is *so* unstable that it simply can't take it anymore. Several neutrons then quickly spit out electrons, turning into protons, so that the element does not just go to the next one up, but takes a giant leap for nuclear-kind, skipping over several element numbers, to make the missing atoms beyond bismuth, including uranium. (Which then proceeds to decay into lead, making radium and a variety of other elements.) The r-process also makes a lot of lighter elements that add to those created by the s-process. Some important ones, including platinum, silver, and gold, are its exclusive domain. It even makes plutonium, element 94. Yes, we make it here for bombs and reactors, but it's still a natural element. Though there is none left on Earth to mine, daughter products from its decay are found in rocks and meteorites.

Rapid-capture elements are again in correct proportion in the Sun. The only problem, the one that makes the r-process mysterious, is that we don't know where it takes place! The most obvious, and the most believed, is in the depths of the exploding core-collapse massive-star supernovae. White dwarf Ia's, though argued, seem to be out. But then:

> If you want to go where the neutrons are,
> You can't get better than a neutron star.

Neutron stars blow off fierce "relativistic" (so fast that relativity must be used to describe them) winds that can power entire supernova remnants, the best known the one in the Crab Nebula. It could be that r-process elements are created *after* the main supernova event within these most intense of all winds.

Whatever the case, when we combine

- Big Bang elements
- elements from energy generation in ordinary stars
- those from the two neutron capture processes
- cosmic ray hits on interstellar atoms
- explosive nuclear reactions in supernovae
- a variety of other reactions

with:

- rates of deposition into the Galaxy
- age of the Galaxy (that is, for how long this has all been going on)

we get something very close to the solar chemistry. And to that of the Earth.

Figure 10.5. Look back in time. Everything within this tiny infrared field of view is a galaxy. The small red patch at center was created only a few hundred million years after the Big Bang. While among the most primitive galaxies ever seen, it nevertheless contains heavy elements made by the first stars. From its point of view, our beautiful mature spiral, the product of billions of years of chemical evolution, would look about the same. NASA, ESA, B. Mobasher (STScI/ESA), and M. Dickinson (Spitzer Space Center).

By looking at the compositions of stars of different ages (as told by a variety of stellar features) we can even see these evolutionary processes at work. Following the zero-metal stars, the first ordinary supernovae had to be of the core-collapse variety. White dwarfs could not be created until the 8 to 10 solar mass stars began to die some 10 to 20 million years after the Galaxy's formation. It would then take far longer for white dwarfs in binaries to overflow their limits and explode as Type Ia's.

The first oxygen and iron to make it into interstellar space to mix in with the raw material for later stellar generations was in the proportion of that produced by core-collapsers. Type Ia's, however, create three times as much iron, and therefore have higher ratios of iron to oxygen. When we look at younger generations of stars, sure enough, the Fe/O ratio goes up! Just what we would expect. To a good extent, everything seems to fit together.

But so what? What does it all mean to *us*?

A Story from Other Worlds

Our Solar System, our planet, which has been the focus of these tales, is not alone. Other systems are everywhere, with other planets that were also made from the leavings of generations of stars. We have now found hundreds of stars with planets, in addition to many multiple-planet systems. We've even seen some planets by direct imaging and through their infrared radiation. However, most make themselves known by gravitationally forcing their parent stars to have small orbits about the systems' centers of mass. They thereby vary the stars' velocities relative to Earth, which are then detected through subtle Doppler shifts. Some planets are also found when they cross in front of their parents, causing the stars to dim slightly. What we are finding now are mostly Jupiter-Saturn-sized bodies, massive planets that produce the greatest effects. But we are also finding some in the Neptune range and some that even begin to approach the mass of Earth: true Earths cannot then be far behind.

Given planets, quite likely other Earths, how about life? We've long looked at other stars for signs of intelligent life. Yet various

Figure 10.6. In another artist's conception, a Jupiter-like planet appears to pass in front of the ninth magnitude star HD 209458, which then backlights its outer layers. The hot planet, carrying seven Jupiter masses, is so close to the star (just 12 percent Mercury's distance from the Sun) that it orbits in a mere 3.5 days. NASA, ESA, and G. Bacon (STScI).

and continuing searches for artificial radio signals have turned up nothing. Are we alone? The great physicist Enrico Fermi (1901–1954) suggested that if the Galaxy contained other civilizations, they would long ago have learned to spread themselves around. They would be everywhere, obvious. But, he asked, where are they? There are many resolutions of the Fermi Paradox, including rarity of Earths, impossibility of interstellar travel, even a "don't care" attitude.

There may be another possibility. Once a sufficient number of planets were listed in the catalogues, a curious relation began to develop. Although there are a number of exceptions, stars with or-biting Jupiter-like planets tend strongly to be more metal-rich than the Sun. The reason is not clear. It may be an effect of selection. The current crop of stars has a bunch of "Jupiters" that lie very close to their parents, the probable result of friction within the planet-forming disks that cause such planets to spiral inward. Metal-rich stars may have thicker disks and thus have "more-discoverable" large bodies that lie close enough to produce good-sized observable effects. But the correlation is likely to be more fundamental. It could also be that stars simply do not create giant planets unless the surrounding disks, from which planets are born, have enough metals in them.

Including its central core, our planet is about a third (by mass) iron and nickel, all created by both kinds of supernovae, as were the silicates that make up the outer mantle and crust. Mercury through Mars have the same ultimate origins, as do the cores of the large planets, as surely do the planets of other stars. If it requires considerable metals to make a planet, and 5 billion years to create a civilization, we could be among the first to have done so. It may have taken 9 billion years (the age of the Universe minus the age of the Sun) for supernovae to make enough heavy materials to beget planets, and ultimately ourselves. True, supernovae go off at a greater rate in the Galaxy's interior, where there are far more stars and the metal abundance reached to appropriate proportions long ago. But the rates of supernovae and of stellar encounters might then be *so* high as to squelch the formation of life in the first place. Countering the concept, the correlation with high metals does not seem to hold for smaller planets, though first, no one knows anything about really small planets like Earth, and second, it is not unlikely that large planets are needed to stabilize a system such as our own.

Speculating with all our might, could we thus be in the vanguard? Did it take this long since the Big Bang to create an intelligent civilization? Are we perhaps the most advanced? Someday, when other societies flourish, will they be looking for us instead of we looking for them?

Heaven's Touch

A starry sky, twinkling lights in a velvet firmament, distant Suns; they enter our arts, our music, poetry, legends, and perhaps most important, our philosophies. But as we vividly see, they also reach out to us to touch us directly in a variety of ways, as does the Moon, the Sun, the other planets in our Sun's System.

Their touch can be destructive, to be sure, but it's also creative. Without it there would be no life at all, indeed, in the grander scheme, there would be no Earth. Look what it took to make us. The Sun was born from a cloud of dusty interstellar gas that was most likely compacted by a supernova shock wave. We in fact find decay products of radioactive elements in meteorites that suggest one went off very close to us as we formed. However, the cloud still had to be cold enough for the contraction to take place. The refrigerator is the dust, which keeps out heating starlight. The dust was made in dying stars: silicates from ordinary ones, carbon-rich from carbon stars, some of it even from supernovae.

Not enough. To condense, the rotation of our birth cloud had to be slowed down. Cosmic rays launched by supernovae caused ionization trails within the cloud that provided hooks for the rotating Galaxy's magnetic field to do the job. The supernovae also gave us the gift of iron, oxygen, silicon, and other elements, without which our own planet could not have been born. And not by small amounts. Given Earth's mass and the number of supernovae that have blown up in our Galaxy before our birth, each one has contributed a mountain-mass to us. We walk on trails made of exploded stars. Neutrinos seem to have no effects, in that Earth is transparent to them. But without them and their outbound pressure in aid of the shockwave from the neutron-rich core, life-giving supernovae might never have occurred in the first place.

Most of our iron comes from white dwarf supernovae. They require stellar duplicity. Without double stars, there would have been no romantically named Type Ia supernovae: and given the need for metals, quite possibly no Earth, and probably no life. Without any of these actions, without nature even making binary stars, we would not be here.

Given a planet, we still need to provide life. The cold conditions in condensing, star-forming clouds create a rich molecular chemistry. We see everything from water and ammonia to complex organics. Some will be destroyed in the furnace of star formation,

Figure 10.7 And here we are back on Earth, our beautiful home, our river of life the product of the stars, the galaxies, and the Universe. J. B. Kaler.

but some may have survived to be added to by complex molecules being made within the disk that generated our planet. It's not unlikely that the seeds of life came to us from outer space, from the disk, then working back, from the stars, from the Galaxy, from the whole Universe, from Heaven's penultimate touch.

✦ Index